Dengzhe Ma
Jürgen Gausemeier
Xiumin Fan
Michael Grafe

Virtual Reality & Augmented Reality in Industry

The 2nd Sino-German Workshop

Dengzhe Ma
Jürgen Gausemeier
Xiumin Fan
Michael Grafe

Virtual Reality & Augmented Reality in Industry

The 2nd Sino-German Workshop

With 161 figures

上海交通大学出版社
SHANGHAI JIAO TONG UNIVERSITY PRESS

Springer

Editors

Dengzhe Ma
Dept. of Mechanical Engineering
Shanghai Jiao Tong University
200030, Shanghai, China
E-mail: dzma@sjtu.edu.cn

Xiumin Fan
Dept. of Mechanical Engineering
Shanghai Jiao Tong University
200030, Shanghai, China
E-mail: xmfan@sjtu.edu.cn

Jürgen Gausemeier
Heinz Nixdorf Institute
University of Paderborn
33102, Paderborn, Germany
E-mail: Jürgen.Gausemeier@ hni.uni-paderborn.de

Michael Grafe
Heinz Nixdorf Institute
University of Paderborn
33102, Paderborn, Germany
E-mail: Michael.Grafe@hni.uni-paderborn.de

ISBN 978-7-313-06426-4
Shanghai Jiao Tong University Press, Shanghai

ISBN 978-3-642-17375-2 e-ISBN 978-3-642-17376-9
Springer Heidebery Dordrecht London New York

Library of Congress Control Number: 2011920985

Printed on acid-free paper

Springer is part of Springer Science+Business Media (www.springer.com)

Preface

Virtual reality and augmented reality (VR/AR) are key technologies for virtual engineering. They are the basis for functional virtual prototyping, which enables engineers to analyse the shape, form and functional behavior of future products in an immersive and interactive virtual environment. Applying these technologies greatly improves the communication in product design and production development: It helps to identify and avoid design errors in early stages of the development process, it reduces the number of physical prototypes and saves time and cost for enterprises. VR/AR are considered as valuable tools for improving and accelerating product and process development in many industrial applications. However, there are still many requirements unaddressed leaving considerable potential for further developing and improving VR/AR-based tools and methods.

This workshop intends to establish an open forum, dedicated to present and discuss innovative applications from industry and research from China and Germany, as well as to exchange experiences. It provides the opportunity to learn about state-of-the-art VR/AR applications in industry, and to directly experience new development and future trends of VR/AR technology.

This workshop is supported by the Department of High and New Technology Development and Industrialization of MOST (Ministry of Science and Technology of China), Deutsches Generalkonsulat Shanghai and Shanghai Science & Technology Committee. The purpose is to promote industrial application of VR/AR tools and methods in China, and to promote the science & technology exchange and cooperation between China and Germany. The publication of these proceedings is also supported by Shanghai JiaoTong University Press and Springer SBM.

We acknowledge all workshop sponsors, organizers and co-sponsors: Shanghai Jiao Tong University, Heinz Nixdorf Institute, University of Paderborn, Shanghai Automotive Industry Science and Technology Development Foundation, Manufacture Information Engineering of China Magazine, Shanghai Academy of Science & Technology, National Engineering Laboratory for Digital Shipbuilding (China), Shanghai Electric Group Co., Ltd. Central Academy, State Key Lab of CAD&CG, State Key Lab of Mechanical System and Vibration (China), Virtual Reality Professional

Committee (China Society of Image and Graphics), Shanghai Science & Technology Museum.

These proceedings can be used as reference for researchers and students in the fields of virtual reality and augmented reality and its application.

During the development of this proceeding, we have received invaluable input and support from the chapter authors, and support from System Simulation and Virtual Reality Group of Shanghai Key Lab of Advanced Manufacturing Environment. We are also grateful to the editors of SJTU press and Springer for their patience and professionalism during the editing process.

Dengzhe Ma
Jürgen Gausemeier
Xiumin Fan
Michael Grafe
Aug. 10th, 2009

Contents

Contributors

Thomas Beer, M. Sc.
Virtual Reality Group, RWTH Aachen University, Seffenter Weg 23, 52074 Aachen, Germany
E-mail: beer@vr.rwth-aachen.de

Johannes Behr, Dr.-Ing.
Fraunhofer-Institut für Graphische Datenverarbeitung, Fraunhofe-rstraße 5, D-64283 Darmstadt, Germany
E-mail: Johannes.Behr@igd.fraunhofer.de

Jan Berssenbrügge, Dr.-Ing.
Heinz Nixdorf Institute, Fürstenallee 11, 33102 Paderborn, Germany
E-mail: Jan.Berssenbruegge@hni.upb.de

Eberhard Bluemel, Dr. rer.nat.
Fraunhofer Institute for Factory Operation and Automation, Sandtorstrasse 22, 39106 Magdeburg, Germany
E-mail: eberhard.bluemel@iff.fraunhofer.de

Ulrich Bockholt, Dr.-Ing.
Fraunhofer-Institut für Graphische Datenverarbeitung, Fraunhofe- rstraße 5, D-64283 Darmstadt, Germany
E-mail: Ulrich.Bockholt@igd.fraunhofer.de

Philippe Cerfontaine, Dipl.-Inform.
Virtual Reality Group, RWTH Aachen University, Seffenter Weg 23, 52074 Aachen, Germany
E-mail: cerfontaine@vr.rwth-aachen.de

Xiumin Fan, Prof. Dr.
CIM Research Institute, Shanghai Key Lab of Advanced Manufacturing Environment, Shanghai Jiao Tong University, HuaShan Road 1954, Shanghai 200030, China
E-mail: xmfan@sjtu.edu.cn

Dieter Fellner, Univ.-Prof. Dr. techn.
Fraunhofer-Institut für Graphische Datenverarbeitung, Fraunhofe-rstraße 5, D-64283 Darmstadt, Germany
E-mail: Dieter.Fellner@gris.tu-darmstadt.de

Jürgen Gausemeier, Prof. Dr.-Ing.
Heinz Nixdorf Institute, Fürstenallee 11, 33102 Paderborn, Germany
E-mail: Juergen.Gausemeier@hni.upb.de

Michael Grafe, Dipl.-Ing.
Heinz Nixdorf Institute, Fürstenallee 11, 33102 Paderborn, Germany
E-mail: Michael.Grafe@hni.upb.de

Bernd Hentschel, M. Sc.
Virtual Reality Group, RWTH Aachen University, Seffenter Weg 23, 52074 Aachen, Germany
E-mail: hentschel@vr.rwth-aachen.de

Hilko Hoffmann, Dr. sc. nat
German Research Center for Artificial Intelligence (DFKI), Stuhlsatzenhausweg 3, Campus D3.2, 66123 Saarbrücken, Germany
E-mail: Hilko. hoffmann@dfki.de

Yong Hu, B.Sc.
CIM Research Institute, Shanghai Jiao Tong University, HuaShan Road 1954, Shanghai 200030, China
E-mail: woodyong@sjtu.edu.cn

Mangqin Jiang, M. Sc.
Design Department, Pan Asia Technical Automotive Center (PATAC)
3999 Longdong Avenue, Pudong District, Shanghai,201201, China
E-mail: Manqing_Jiang@patac.com.cn

Yicheng Jin, Prof.
Key Lab. of Marine Simulation & Control, Dalian Maritime University Dalian 116026, China
E-mail: jycdmu@dlmu. edu.cn

Sascha Kahl, Dipl.-Wirt.-Ing.
Heinz Nixdorf Institute, Fürstenallee 11, 33102 Paderborn, Germany
E-mail: Sascha.Kahl@hni.upb.de

Torsten Kuhlen, Prof. Dr. rer. nat.
Virtual Reality Group, RWTH Aachen University, Seffenter Weg 23, 52074 Aachen, Germany
E-mail: kuhlen@vr.rwth-aachen.de

Xiuwen Liu, Dr.
Key Lab. of Marine Simulation & Control, Dalian Maritime University Dalian 116026, China
E-mail: Liuxw@dlmu.edu. cn

Alexander Löffler, M. Sc.
Computer Graphics Group, Saarland University (UdS), Stuhlsatzenhausweg 3, Campus D3.2, 66123 Saarbrücken, Germany

Dengzhe Ma, Prof. Dr.-Ing.
CIM Research Institute, Shanghai Key Lab of Advanced Manufacturing Environment, Shanghai Jiao Tong University, HuaShan Road 1954, Shanghai 200030, China
E-mail: dzma@sjtu.edu.cn

Zhigeng Pan, Prof. Dr.
State Key Lab of CAD&CG, Zhejiang University, Hangzhou, 310027, China
E-mail: zgpan@cad.zju.edu.cn

Michael Repplinger, Dipl.-Inform
German Research Center for Artificial Intelligence (DFKI)
Stuhlsatzenhausweg 3, Campus D3.2, 66123 Saarbrücken, Germany
E-mail: michael.repplinger@dfki.de

Hongxiang Ren, Dr.
Key Lab. of Marine Simulation & Control, Dalian Maritime University Dalian 116026, China
E-mail: dmu_rhx@163.com

Junsheng Ren, Dr.
Key Lab. of Marine Simulation & Control, Dalian Maritime University Dalian 116026, China
E-mail: j.s.ren@126.com

Jürgen Roßmann, Prof. Dr.-Ing.
Institute of Man-Machine-Interaction, RWTH Aachen University, Ahornstr. 55, 52074 Aachen, Germany
E-mail: rossmann@mmi.rwth-aachen.de

Dmitri Rubinstein, Dipl.-Inform
German Research Center for Artificial Intelligence (DFKI)
Stuhlsatzenhausweg 3, Campus D3.2, 66123 Saarbrücken, Germany
E-mail: dmitri.rubinstein@dfki.de

Michael Schenk, Prof. Dr. rer.nat.
Fraunhofer Institute for Factory Operation and Automation, Sandtorstrasse 22, 39106 Magdeburg, Germany
E-mail: michael.schenk@iff.fraunhofer.de

Marco Schumann, Dipl.-Inform
Fraunhofer Institute for Factory Operation and Automation, Sandtorstrasse 22, 39106 Magdeburg, Germany
E-mail: marco.schumann@iff.fraunhofer.de

Philipp Slusallek, Prof.Dr.-Ing.
German Research Center for Artificial Intelligence (DFKI)
Stuhlsatzenhausweg 3, Campus D3.2, 66123 Saarbrücken, Germany
E-mail: philipp.slusallek@dfki.de

Frank Thielemann, Dr.-Ing.
UNITY AG, Lindberghring 1, 33142 Bueren, Germany
E-mail: frank.thielemann@unity.de

Helene Wassmann, Dipl.-Inform.
Heinz Nixdorf Institute, Fürstenallee 11, 33102 Paderborn, Germany
E-mail: Helene.Wassmann@hni.upb.de

Andreas Wierse, Dr.-Ing.
VISENSO GmbH, Nobelstraße 15, 70569 Stuttgart, Germany
E-mail: aw@visenso.de

Marc Wolter, Dipl.-Inform.
Virtual Reality Group, RWTH Aachen University, Seffenter Weg 23, 52074 Aachen, Germany
E-mail: wolter@vr.rwth-aachen.de

Dianliang Wu, Dr.
CIM Research Institute, Shanghai Key Lab of Advanced Manufacturing Environment, Shanghai Jiao Tong University, HuaShan Road 1954, Shanghai 200030, China
E-mail: dlwu@sjtu.edu.cn

Rundang Yang, Dr.
CIM Research Institute, Shanghai Jiao Tong University, HuaShan Road 1954, Shanghai 200030, China
E-mail: yrd@sjtu.edu.cn

Yong Yin, Prof. Dr.
Key Lab. of Marine Simulation & Control, Dalian Maritime University Dalian 116026, China
E-mail: bushyin_dmu@263.net

Ruwei Yun, Associate Prof. Dr.
Edu-Game Center, Nanjing Normal University, Nanjing, 210097 China
E-mail: yunruwei@njnu.edu.cn

Martin Zimmermann
VISENSO GmbH, Nobelstraße 15, 70569 Stuttgart, Germany
E-mail: mz@visenso.de

Xiufeng Zhang, Dr.
Key Lab. of Marine Simulation & Control, Dalian Maritime University Dalian 116026, China
E-mail: zxfdmu@163.com

Xijin Zhen, Dr.
CIM Research Institute, Shanghai Jiao Tong University, Huashan Road 1954, Shanghai 200030, China
E-mail: zhenxj@sjtu.edu.cn

Hongmin Zhu, B.Sc.
CIM Research Institute, Shanghai Jiao Tong University, Huashan Road 1954, Shanghai 200030, China
E-mail: zhm_810304@sjtu.edu.cn

Design and VR/AR-based Testing of Advanced Mechatronic Systems

Jürgen Gausemeier, Jan Berssenbrügge, Michael Grafe, Sascha Kahl and Helene Wassmann
Heinz Nixdorf Institute, University of Paderborn, Germany

Abstract

Advanced mechatronic systems with inherent partial intelligence, so-called self-optimizing systems, react autonomously and flexibly on changing environmental conditions. Such systems are capable of learning and optimizing their behavior during operation. Their principle solution represents a significant milestone because it is the result of the conceptual design as well as the basis for the concretization of the system itself, which involves experts from several domains, such as mechanics, electrical engineering/electronics, control engineering and software engineering. Today, there is no established design methodology for the design of advanced mechatronic systems. This contribution presents a new specification technique for the conceptual design of advanced mechatronic systems along with a new approach to manage the development process of such systems. We use railway technology as a complex example to demonstrate, how to use this specification technique and to what extent it facilitates the development of future mechanical engineering systems. Based on selected virtual prototypes and test beds of the RailCab we demonstrate, how VR- and AR-based approaches for a visual analysis facilitate a targeted testing of the prototypes.

Keywords

Mechatronics, Self-Optimization, Design Methodology, Principle Solution, Targeted Testing, Virtual Prototype, Visual Analysis, Virtual / Augmented Reality

1 Virtual Prototyping in the Product Innovation Process

Products and manufacturing systems of mechanical engineering and its related industrial sectors like automotive engineering are getting more and more complex. Time-to-market is decreasing simultaneously. Under these circumstances the product innovation process is facing extraordinary challenges. Before we point out how to overcome these challenges, let us spend a brief look on the product innovation process.

The product innovation process starts from the idea of a product or business and leads to the successful product launch. It incorporates the areas of product planning, R&D and manufacturing process planning. The general work flow is shown in the figure. In practice, the product innovation process is iterative and comprises a number of cycles (see Fig. 1).

The first cycle characterizes the steps from finding the success potentials of the future to creating the promising product design, what we call the principle solution. There are four major tasks in this cycle:

- foresight
- product discovering
- business planning
- conceptual design

The aim of **foresight** is to recognize the potentials for future success, as well as the relevant business options. We use methods such as the scenario technique, Delphi studies and trend analysis.

The objective of **product discovering** is to find new product ideas. We apply in this phase creativity techniques such as the Lateral Thinking of de Bono or the well-known TRIZ.

Business planning is the final task in the cycle of strategic product planning. It initially deals with the business strategy, i.e. answering the question as to which market segments should be covered, when and how. The product strategy is then elaborated on this basis. This contains information:

- on setting out the product program
- on cost-effectively handling the large number of variants required by the market
- on the technologies used and
- on updating the program throughout the product lifecycle

Additionally, a business plan must be worked out to make sure an attractive return on investment can be achieved.

This first cycle is also concerned with the **conceptual design,** although this area of activity is actually assigned to product development in the strict sense. The result of the conceptual design is the principle solution. It is, for example, required to determine the manufacturing costs needed in the business plan. That is the reason why there is a close interaction between strategic product planning and product design linked by conceptual design. Conceptual design is the starting point for the next cycle.

This second cycle corresponds to the established understanding of product development. The essential point here is the refinement of the cross-domain principle solution by the domain experts involved, such as mechanical engineering, control technology, electronics and software engineering. The results elaborated by the domains in this cycle must be integrated into an encompassing product specification. This specification has to be verified in the light of the requirements given by the first cycle.

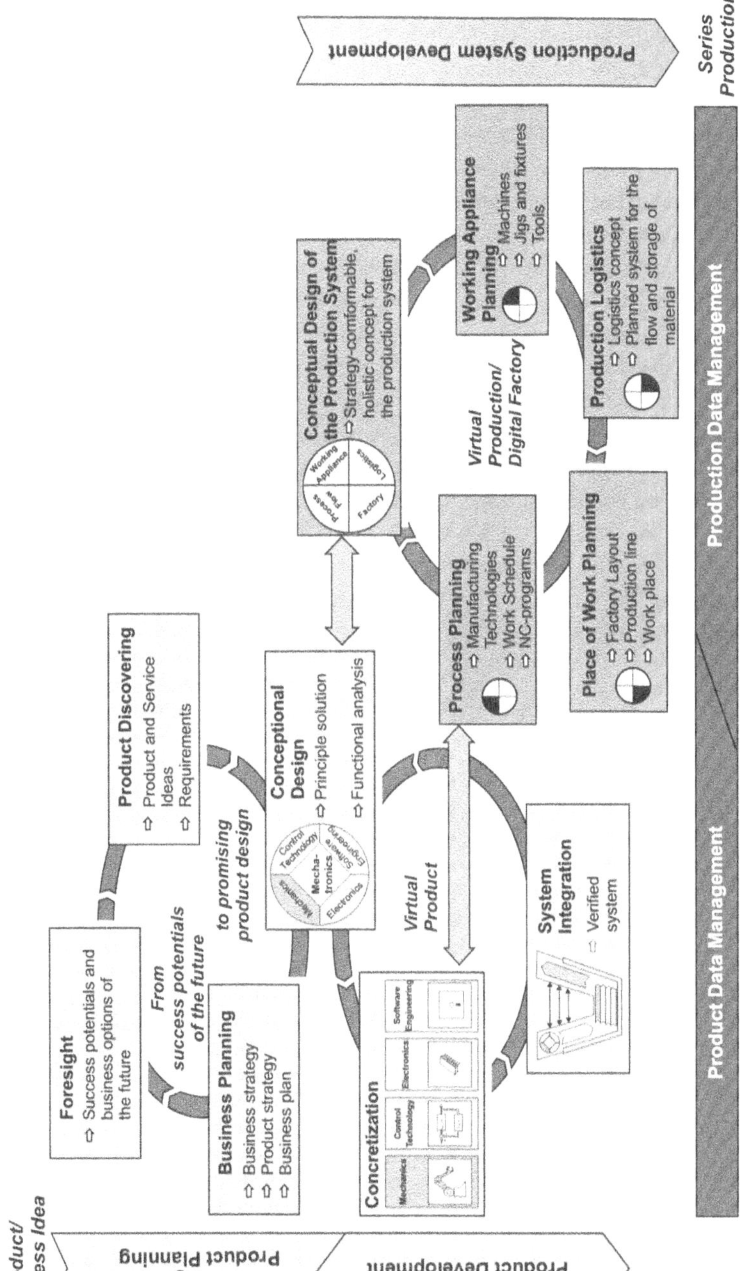

Fig. 1 The product development process as a sequence of cycles [1]

This is done in the product integration phase.

The third and last cycle focuses on manufacturing process development and the optimization of the product design with respect to manufacturing.

The second and the third cycle Product Development and Development of the corresponding manufacturing process are decisively driven by information technology. A key element is Virtual Prototyping. It means to build and analyze computer models of products and production systems being developed in order to reduce time- and cost-intensive manufacturing and testing of prototypes to a minimum. Simulation is another term for experimenting with such computer models. When we model products in the computer, we talk about the virtual product; in analogy we use the buzzwords virtual production or digital factory when we model the manufacturing system.

A perfect virtual prototype represents all aspects of a product (see Fig. 2). 3D-CAD systems are basically used to model the shape of parts. The breakdown of the product to its parts and assemblies is represented by the product structure. Therefore, it is necessary to set up a Product Data Management (PDM). The shape of individual parts in conjunction with product structure is used to develop a shape-based design of the product, what we call Digital Mock Up (DMU). It represents the spatial composition of all parts and assemblies of the product. A DMU can be used to carry out experiments such as clash detection, checking assembly and disassembly sequences. This is all based on the shape of the technical system. To analyze the behavior, we need to consider additional aspects offered by a virtual prototype.

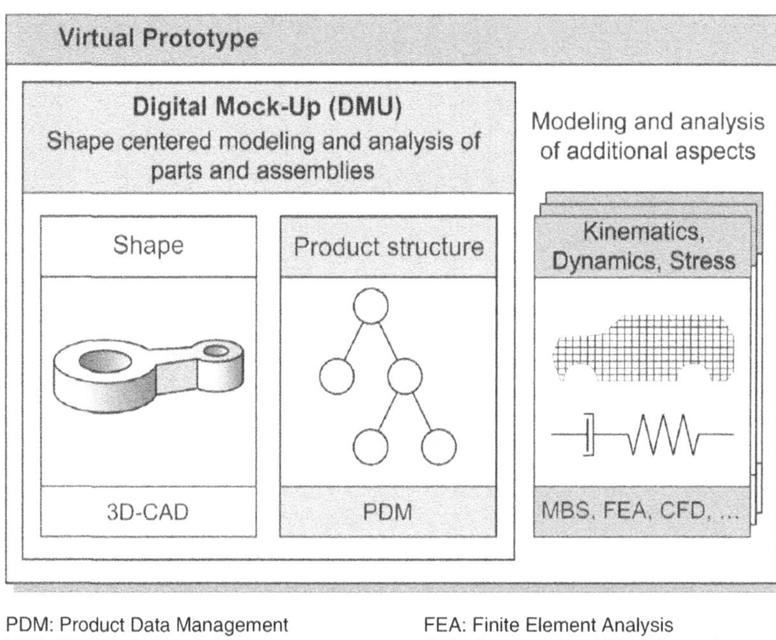

Fig. 2 From solid modeling to virtual prototyping [2]

As Fig. 2 illustrates, we consider a virtual prototype as an extension to DMU since it covers additional aspects such as kinematics, dynamics, and stress. A virtual prototype represents not only shape but also functional features and behaviors. Although, virtual prototyping cannot completely replace experiments with real prototypes, it scientifically contributes to a shorter time-to-market and less development costs even for more complex products and production systems.

2 Virtual Reality (VR) and Augmented Reality (AR)

Virtual Reality (VR) and Augmented Reality (AR) are key technologies of Virtual Prototyping. They are easy-to-understand user interfaces to a virtual design space and facilitate an interactive exploration of the functionality of a new product.

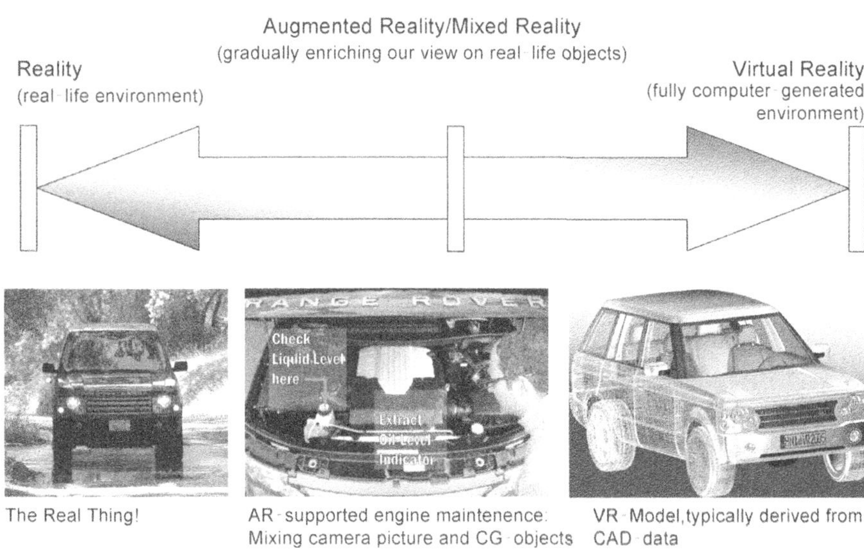

Fig. 3 From real-life to virtual reality

VR means a fully computer generated, three-dimensional environment, in which the engineer can interact with and manipulate a realistic representation of the product in real time. AR goes one step beyond: in contrast to VR, AR enriches the user's view on the real world with virtual objects, which are placed at right time and position regarding the user's perspective. Figure 3 shows an example for the transition from Reality (the real Range Rover) to AR (mixing the real engine with computer generated object for maintenance purposes) to VR (a realistic 3D-modell of the car). In details, VR technology can be characterized by the following main aspects:

Firstly, VR stands for a **realistic rendering** of the product appearance (material, surface, colors) and behavior. Secondly, VR makes use of **advanced display**

technologies that allow the engineers to experience the virtual prototype like a real one. Figure 4 (top left) shows some typical VR-display systems: Head mounted displays (HMD) with small LCD-monitors in front of the eyes. They allow a spatial view on the virtual prototype, but offer less image quality and wearing comfort.

In consequence, today most industrial VR applications use projection-based display systems that consist of several projections in different configurations. The typical configurations of such systems include PowerWall (Fig. 4, lower half) for group presentation or CAVE (Fig. 4, top right) for more spatial immersion of the users.

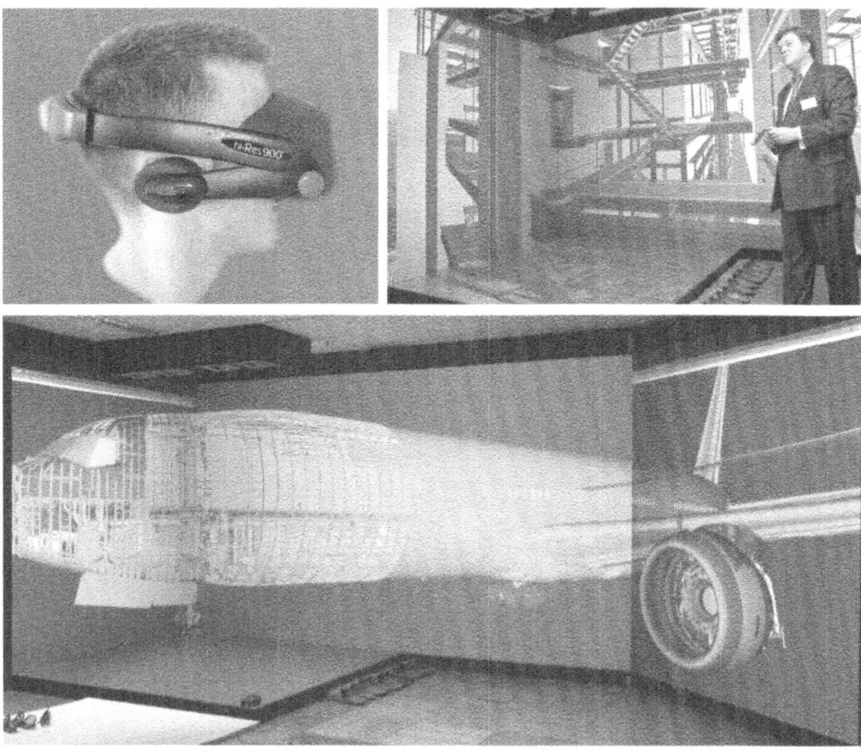

Fig. 4 Sample display devices used for VR applications: HiRes900 HMD (top left, source: Daeyang), CAVE application (top right, source: HD-Visualisation Center, HNI) and powerwall system at HNI HD-Visualisation Center (lower half, source for visualized dataset: Boeing)

In VR, the engineer has to navigate in 3D-space and to manipulate 3D-objects. Therefore, **VR-specific devices for spatial interaction** like 3D-Mouse, 3D-Wands or gloves are needed. By the help of 3D-position tracking systems, the VR system knows the position and orientation of the user in the virtual environment and is able to interpret the navigation and manipulation commands.

The main challenge of AR technology is the **context-sensitive mixture of real world elements and computer generated objects in the user's field of view**. Therefore, an exact position tracking of the user inside the real world is needed in real-time. Based on this information, the AR-system determines the virtual objects to be shown, their size and position in the user's field of view.

Today there are two different approaches to display the mixed image of real and virtual objects: "Video-see-through" display devices have integrated miniature video cameras. The user sees a real-time video stream of his real environment, which is enriched with computer-generated objects. Figure 5 (top right) shows a video-see-through HMD published by Canon in 2002 and its application in car door assembly (see Fig. 5, top left).

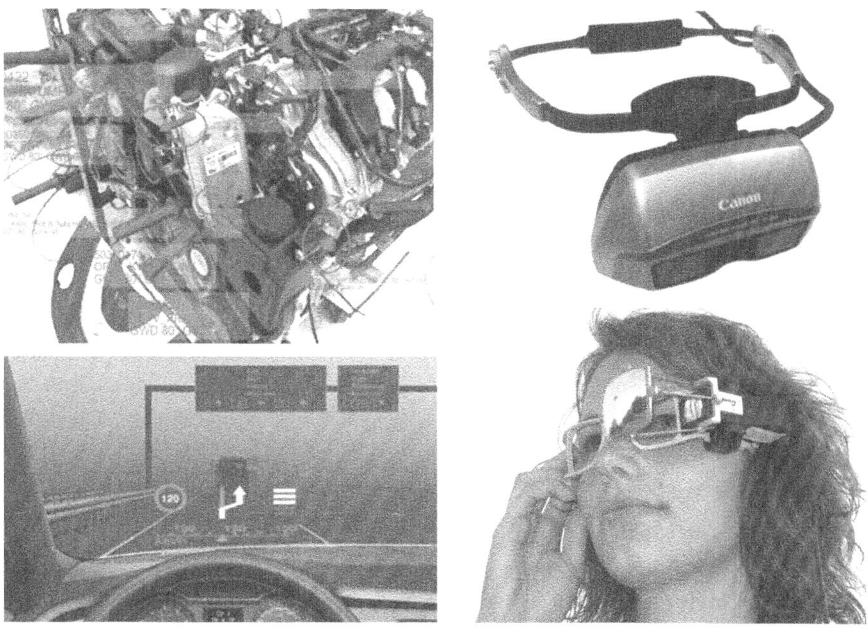

Fig. 5 Sample display devices and AR applications: AR-system for vehicle assembly (top left, source: Brose), video-see-through HMD (top right, source: Canon), automotive head-up display (lower left, source: AUDI) and optical-see-through glasses (lower right, source: Zeiss)

In Fig. 5 (lower left), we see an example of "optical-see-through" display in a car. The computer generated objects are directly projected in the front window. Here, the complex and time-consuming video processing is not necessary. Figure 5 (lower right) shows a high resolution optical-see-through HMD from Zeiss.

3 Up-to-date Applications of VR and AR in Industry

In recent years, VR technology has successfully made its way from research institutes into industrial practice. VR today is a key technology in industry, e.g. in product development, plant engineering and service. It facilitates the engineer's understanding of complex design concepts and allows more efficient interaction between the engineer and the computer. This saves time and money and finally enhances product quality. Augmented Reality is still at the beginning of its industrial employment, however first joint research projects with partners from industry show the great benefit of this fascinating technology.

The following projects of the Heinz Nixdorf Institut give a survey about how VR and AR technology could be applied in the product innovation process.

3.1 Composing Mechatronic Systems in VR

The complexity of modern mechatronic systems and the necessity to efficiently analyze and explore their large number of potential configurations and behavior patterns ask for new development methods and tools. We developed a virtual prototyping environment[1], which allows the engineers to interactively compose mechatronic prototypes in the virtual world [3,4]. The design approach we used is based on the combined effects of interconnected system elements, also referred to as "solution elements", like sensors, actuators, and mechanical parts as well as Mechatronic Function Modules (MFM).

During the design in the virtual environment, the system composition is executed in a synthesis and analysis cycle (Fig. 6). In the first step, the system synthesis, the engineer interactively arranges the assembly structure. Therefore, he uses 3D-models of solution elements from a construction library. In parallel, the active structure of the assembled system will be automatically deduced by the design system. The active

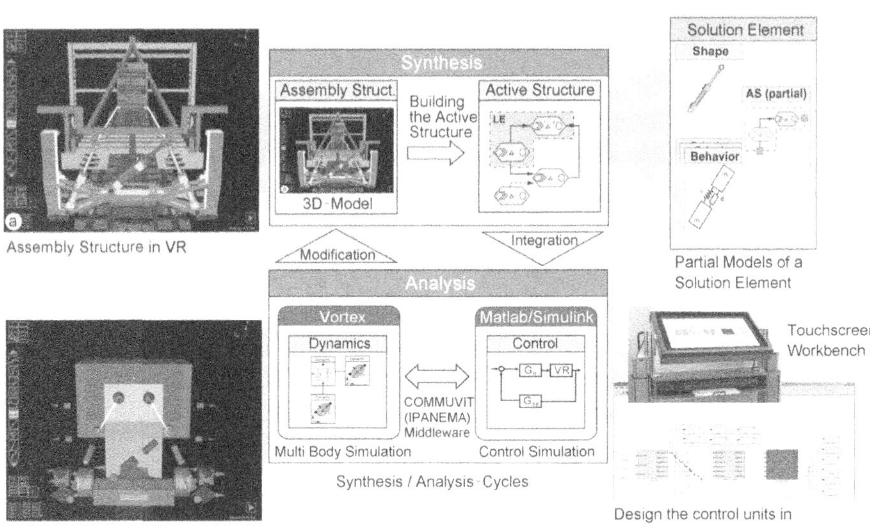

Fig. 6 Synthesis / analysis cycle in the composition of mechatronic systems

[1] This research was conducted as a part of the Collaborative Research Center 614 "Self-Optimizing Concepts and Structures in Mechanical Engineering", which is supported by German Research Foundation.

structure comprises all relevant configurations of solution elements (controller, sensors, and actuators) and their interrelationships. In the second step, the engineer can analyse the dynamic behaviour of the designed system in real-time. During system analysis Matlab/Simulink is used for the simulation of the system control. The simulation of the multi body system is done by VORTEX.

In order to make the interaction of system composition more intuitive, a set of gestures are defined as commands for the engineer to execute the assembly actions in VR (Fig. 7).

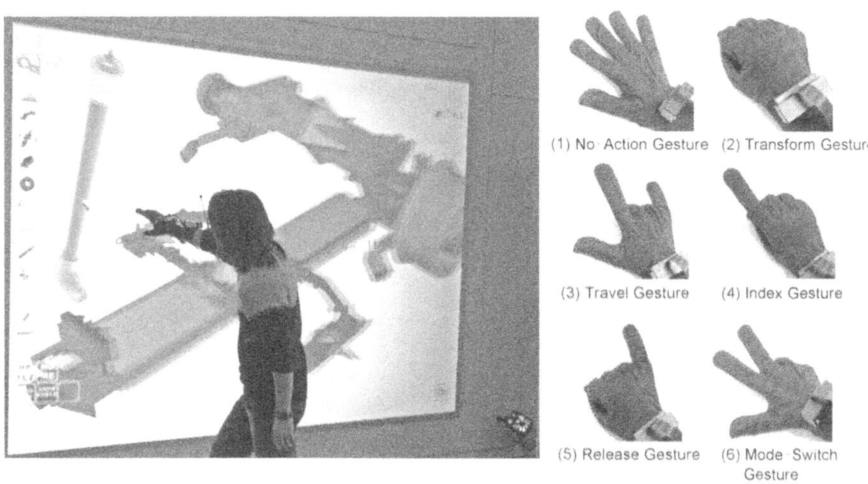

Fig. 7 Composing a virtual mechatronic prototype via intuitive gestures

To facilitate an intuitive exploration of the virtual prototype, various visual effects (see Fig. 8) were applied in the working environment. For example, animations are used to show dynamics behavior; 3D-annotations, like the strengths and directions of forces, present the simulation results more comprehensively and are easier to be observed in the virtual environment. The system enables the engineer to get a first understanding of the behavior of the mechatronic system at a very early stage of the development process. Design errors could be detected; design alternatives could be intuitively tested without the need of a real prototype.

Fig. 8 Visual representation for simulation and analysis

3.2 Virtual Prototyping of Headlight Systems

Modern automobiles contain more and more mechatronic components to support the task of driving. Such mechatronic components are, e.g., an anti-lock braking system (ABS) and an electronic stability program (ESP) to support driving safety, or advanced front lighting system (AFS) to enhance the lighting capabilities of a vehicle on a winding road.

Dynamic bending lights typically use the steering wheel angle or a gyro sensor to calculate the swivelling angle of the headlights to enhance vehicle lighting on a bending road. These systems provide a headlight that "follows" the course of a winding road and are referred to as advanced front lighting systems (AFS) (Fig. 9). The main disadvantage of such systems is that the vehicle lighting heads into a curve too late.

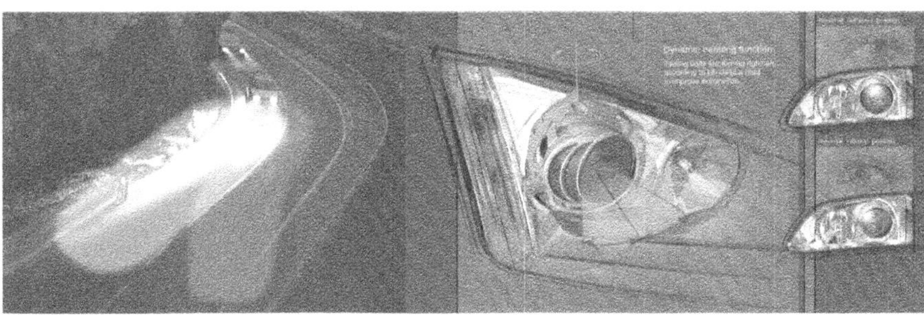

Fig. 9 Advanced Front Lighting (AFS) (Source: Visteon)

Therefore, our project partner Visteon, a global supplier to the automotive industry, has developed a predictive advanced front lighting system (P-AFS) [5]. P-AFS uses GPS-data to locate the vehicle's position plus digital map data to predict the curvature of the road in front of the vehicle. Based on this, P-AFS predicts the road scenario and turns the front headlights accordingly. That way, the headlights follow the road's curvature and optimally illuminate the road in front of the vehicle.

In design of headlight systems, we need to do numerous tests for the different headlight. In order to reduce the time and cost spent on the tests, we developed Virtual Night Drive (VND) (Fig. 10), a PC-based night drive simulator [6]. VND utilizes the Shader technology of latest generation graphics systems to visualize the complex lighting characteristics of modern automotive headlight systems in high detail and in real-time during a simulated night drive [7]. The user of VND drives a simulated vehicle on a virtual test track. Different headlight systems are supported and the user can choose between various headlamp models and display modes.

The motion of the vehicle, resulting from the vehicle dynamics and the user's driving action, directly influences the lighting direction of the vehicle's headlights. The effect on the illumination of the road in front of the vehicle is visualized in real-time and can be evaluated on screen immediately.

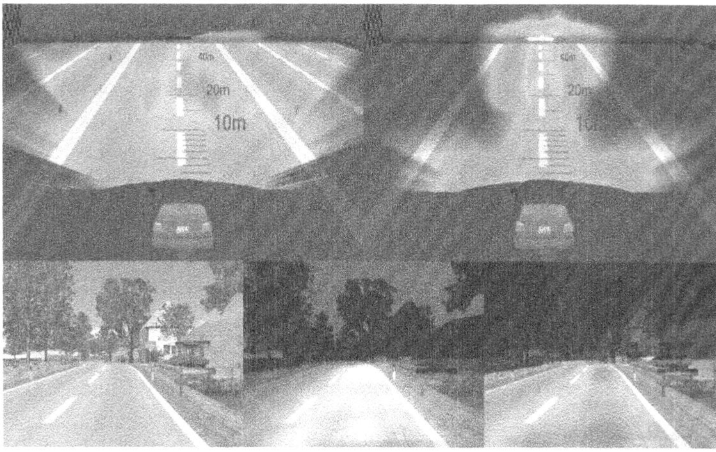

Fig. 10 Real-time visualization in Virtual Night Drive: low beam (upper left), high beam (upper right), and display of day light, night time, and false color modes (bottom half)

The system can be scaled from a single-PC version with one screen and simple steering wheel up to several linked PCs firing a multi-channel stereo display system. Each PC drives one channel of the projection system in mono or stereo mode. A complete cockpit of a small vehicle (Smart-for-two by Daimler Chrysler) is connected to the system as the IO-interface to the user (see Fig. 11). Several test-setups are installed on site at project partners both from academia and industry.

Fig. 11 Flexible presentation setup for VND: scaling VND from simple desktop-version to full-scale simulator application

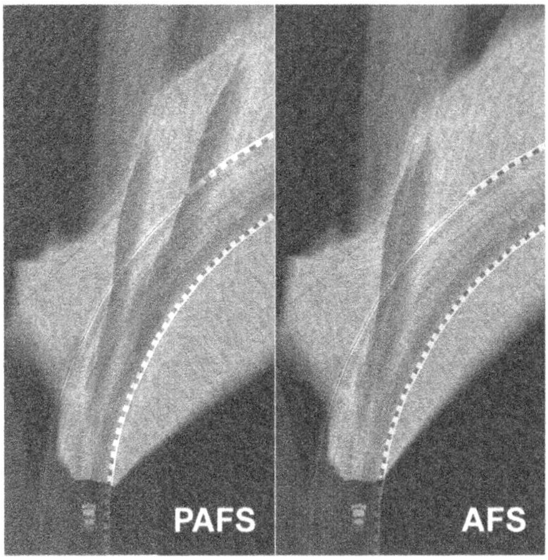

Fig. 12 Evaluation of PAFS within VND: bird's eye view shows early swiveling of PAFS (left) and conventional AFS (right)

The prototypic integration into our VND-system simulates both predictive and conventional AFS and facilitates a direct comparison of both systems on screen. To illustrate the functional behavior of both systems more clearly, different viewing position can be chosen interactively, while the system is running. Bird's eye view (Fig. 12) reveals the different timings between P-AFS and conventional AFS when turning the headlights to follow the oncoming bend of the road.

3.3 AR in Vehicle Prototyping

Currently, the ergonomic design becomes an essential factor in judging the success of a car model. In particular, the increasing features for security, comfort and communication require conclusive view and control concepts to reduce the stress for the driver. For a long time, real prototypes have been used in the conceptual design of vehicles. Physical, 1:1-scale mock-ups of a driver's cabin with seats and dashboard allow the engineer to evaluate the interior design (Fig. 13 left). However, to build or modify such real prototypes is time consuming and expensive. In addition, commercial CAx software based on 3D-modells of the car and the driver facilitate a basic analysis of the ergonomic design (Fig. 13 right). They allow e.g. the calculation of the passenger's leg room or his field of view. Besides those computational methods, the quality of vehicle ergonomics is heavily influenced by perception of the real passenger. For this, the physical mock-up is still in use.

In recent years, Volkswagen Commercial Vehicles developed a VR-based virtual

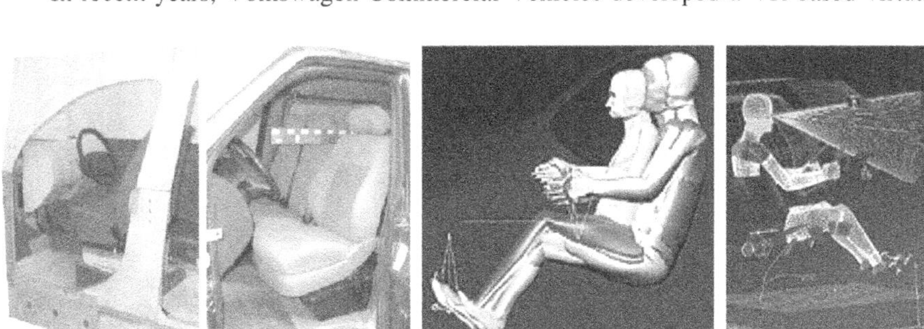

Fig. 13 Physical Mock-Up of the driver's cabin in 1:1-scale (left) and Analysis of vehicle ergonomics using 3D human models (right, Source: Volkswagen Commercial Vehicles AG)

prototyping system (Fig. 14). The test person sits in a real driving seat and wears a head mounted display. The person feels itself located in a virtual car, in which the person can steer through a virtual city. This allows an intuitive analysis of visibility conditions concerning e.g. the instrument panel, traffic lights or pedestrians.

Fig. 14 VR driving seat (Source: Volkswagen Commercial Vehicles AG)

Compared to the physical models, this system has been a step ahead. But the realistic representation of the virtual environment needs a lot of effort, especially the simulation of the vehicle dynamics cost a lot of modeling time and computational power.

In a joint research project together with Volkswagen Commercial Vehicles we thought about a new approach: Why not combine a real car with virtual car components to get best of both worlds? Therefore, we developed an AR-based mobile test bench for vehicle ergonomics [8].

The basic platform of the mobile test bench is a real car without dashboard, rear seats, columns and ceiling. The driver is seated in his normal driving position with a HMD on. The HMD has a pair of integrated video cameras and LC-displays. The cameras grab the driver's view on the real world. The position and viewing direction of the driver are captured by a high precision optical tracking system (Fig. 15 left). While driving the mobile test bench, the driver sees the real car and test track augmented with virtual car components (Fig. 15 right).

Fig. 15 Mobile AR test bench with tracking system for position detection (left) and virtual car components in driver's field of view (right)

This allows an intuitive, fast check of the visibility condition inside and outside the car. Design variants could be easily tested by a simply exchange of the virtual components. In comparison to former approaches, the mobile test bench is much more flexible and less time consuming. The benefit has been approved by several case studies together with engineers from Volkswagen. Now, the project has been extended to integrate further functions like high quality rendering or interactive manipulation of the virtual components. Within the next months, Volkswagen intends to apply the mobile AR test bench in a new car development project.

4 From Mechatronics to Self-Optimization

The term "mechatronics" stands for the symbiotic cooperation of mechanics, electronics, control engineering and software engineering. That opens fascinating perspectives for the development of future mechanical engineering products. The variety of mechatronic systems can be expressed by two categories:

- The first category is based on the spatial integration of mechanics and electronics. The aim is to reach a high density of mechanical and electronic functions within available space. The focus of this class lies on assembly and connecting technologies e.g. Molded Interconnect Devices (MID).
- The second category deals with the controlled movements of multi-body systems. Here sensors collect information about the environment and the system itself. The control system utilizes this information to derive appropriate reactions. These reactions are then executed by the system's actuators. That is what we call the basic control loop of a mechatronic system.

A representative example for the second category of mechatronic systems is the project "RailCab". Here the control technology is in the focus of attention. RailCab is an innovative rail system that has been realized as a test facility on a scale of 1:2.5 at the University of Paderborn (www.railcab.de). The core of the system comprises autonomous vehicles (shuttles) for transporting passengers and goods according to individual demands rather than a timetable (Fig. 16). They act proactively, e.g. they form convoys to increase capacity utilization and reduce energy consumption. The shuttles are driven by an electromagnetic linear drive. The main part of the shuttle's technics is located in its flat floor pan to which the different cabins for passengers and cargo are attached. Figure 16 shows the shuttle itself and the system's capability to form convoys automatically.

Usually mechatronic systems of this category handle more than one control task, which have to be coordinated. To cope with this complexity our colleague J. Lückel has introduced a hierarchical structure of three levels [9] (Fig. 17).

The basis of this is provided by so called "mechatronic function modules", consisting of a basic mechanical structure, sensors, actuators and a local information processor containing the controller.

A combination of mechatronic function modules, coupled by information

Fig. 16 RailCabs of the project "Neue Bahntechnik Paderborn/RailCab"

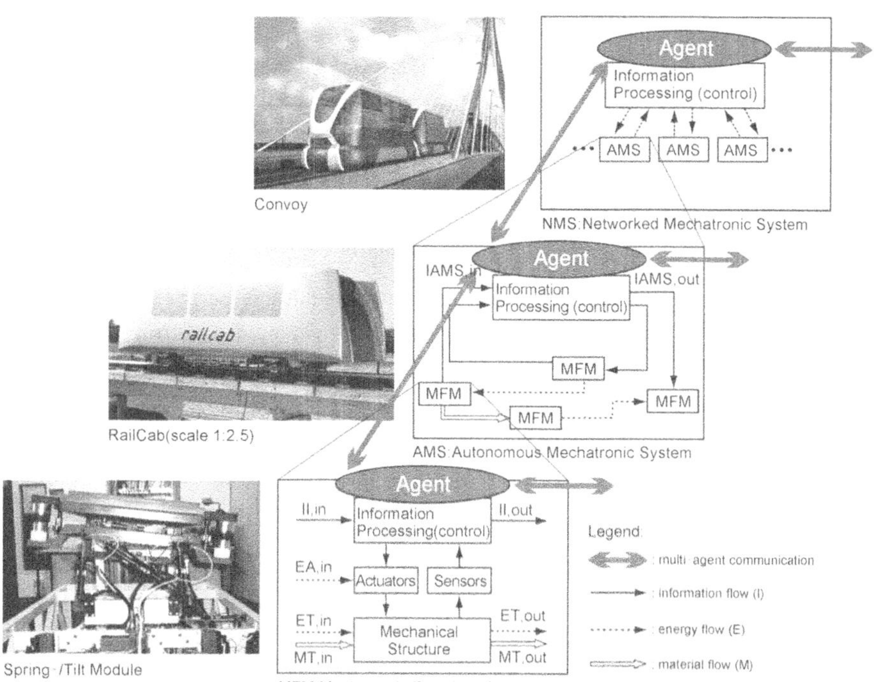

Fig. 17 Hierarchical structure of a self-optimizing system

technology and mechanical elements, constitutes an "autonomous mechatronic system". Such systems also possess a controller, which deals with higher-level tasks such as monitoring, fault diagnosis and maintenance decisions as well as generating parameters for the subordinated information processing systems of the mechatronic function modules.

Similarly, a number of autonomous mechatronic systems constitute a so called "networked mechatronic system", simply by coupling the associated autonomous mechatronic systems via information processing.

In the context of vehicle technology, a spring and tilt module would be a mechatronic function module, the shuttle would be an autonomous mechatronic system, and a convoy would be a networked mechatronic system. Because of the high complexity and the participation of different domains the development of mechatronic systems is still a challenge, let alone of self-optimizing systems.

We use the term "self-optimization of a technical system" to describe the endogenous adaptation of the system's objectives to changing environmental conditions and the resultant autonomous adaptation of its behavior. Self-optimization facilitates systems with inherent "intelligence" that are able to take action and react autonomously and flexibly to changing operating conditions.

Reverting to the hierarchical structure of a mechatronic system (Fig. 17) on each level the controllers are enhanced by the functionality of self-optimization. Thus, the previously mentioned modules and systems (that means mechatronic function modules, autonomous mechatronic systems and networked mechatronic systems) receive an inherent partial intelligence.

The key aspects and the mode of operation of a self-optimizing system are depicted in Fig. 18. Using the influences as a basis, the self-optimizing system determines the internal objectives that have to be pursued actively. These internal objectives are based on external ones, whereas those are set from the outside, e.g. by the user or other systems, and also on inherent objectives that reflect the design purpose of the system. Inherent objectives of a driving module can be for example: ensuring the driving functions and a high efficiency. When we subsequently talk about objectives, we refer to the internal ones, because those are part of the optimization. Low energy demand, high travelling comfort and low noise emission belong to internal objectives of a shuttle. The adaptation of objectives means, for instance, that the relative weighting of the objectives is modified, new objectives are added or existing objectives are discarded and no longer pursued.

Thus, the adaptation of the objectives leads to an adaptation of the system behavior. The behavior's adaptation is achieved by an adaptation of the parameters and, if necessary, of the structure itself. An adaptation of the parameters means an adaptation of the system's parameters, e.g. the adaptation of a controller parameter.

Adapting the structure, concerns the arrangement and relations of the system's elements. We differentiate between reconfiguration and compositional adaptation. It is reconfiguration, if the relations of a fixed quantity of elements are changed. Compositional adaptation means the integration of new elements in the already existing

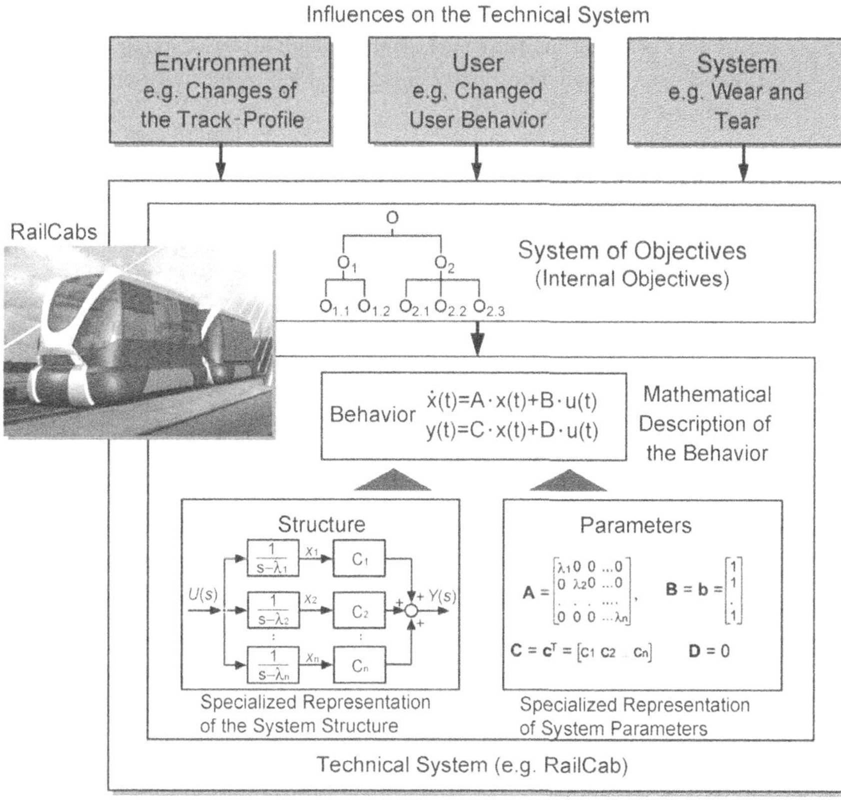

Fig. 18 Key aspects of a self-optimizing system

structure or the subtraction of elements from the structure. Self-optimization takes place as a process that consists of the three following actions, called the **Self-Optimization Process:**

(1) **Analyzing the current situation:** The regarded current situation includes the current state of the system as well as all observations of the environment that have been carried out. Observations can also be made indirectly by communication with other systems. Furthermore, a system's state contains possible previous observations that are saved. One basic aspect of this first step is the analysis of the fulfillment of the objectives.

(2) **Determining the system's objectives:** The system's objectives can be extracted from choice, adjustment and generation. By choice we understand the selection of one alternative out of predetermined, discrete, finite quantity of possible objectives; whereas the adjustment of objectives means the gradual modification of existing objectives respectively of their relative weighting. We talk about generation, if new objectives are being created that are independent from the existing ones.

(3) **Adapting the system's behavior:** The changed system of objectives demands

an adaptation of the behavior of the system. As mentioned before this can be realized by adapting the parameters and, if required, by adapting the structure of the system. This action finally closes the loop of the self-optimization by adapting the system's behavior.

The self-optimizing process leads, according to changing influences, to a new state. Thus, a state transition takes place. The self-optimizing process describes the system's adaptive behavior. This can occur on every hierarchy level of a self-optimizing system shown in Fig. 17. The realization of complex, mechatronic systems with inherent partial intelligence requires an adequate concept of structure as well as architecture for the information processing. To make this possible, a new concept has been developed: **Operator-Controller-Module (OCM)** [10]. From an information processing point of view, it corresponds to an agent.

5 Design of Advanced Mechatronic Systems

The development of advanced mechatronic systems is still a challenge. The established design methodologies, i.e. the engineering design by Pahl/ Beitz [11] or the VDI Guideline 2206 [12], lay the foundation to meet these challenges. Nevertheless, these methodologies need to be fundamentally extended and added by domain-spanning methods and tools to handle the complexity of the development. This especially applies to the early development phase "conceptual design".

On the highest degree of abstraction, the development process of advanced mechatronic systems can be subdivided into the domain-spanning conceptual design and the domain-specific "concretization" (see Fig. 19). Within the conceptual design, the basic structure and the operation mode of the system are defined. Thus, the conceptual design has to include the decomposition of the system into modules. The decomposition results into a development-oriented product structure, which integrates the two basic and mostly contradictious views of shape- and function-oriented structure. All results of the conceptual design are specified in the so-called "principle solution". A set of specification techniques, in order to describe the principle solution of advanced mechatronic systems, has been developed. By using this specification technique, the system that is to be developed will be described in a holistic, domain-spanning way. The description of the principle solution provides all relevant information for the structuring of the system and forms the basis for the communication and cooperation of the developers from different domains. Based upon the principle solution the subsequent domain-specific "concretization" is planned and realized. The term "concretization" describes the domain-specific design of a technical system, based on the principle solution. The aim of the concretization is the complete description of the system by using the construction structure and the component structure. In so doing, all defined modules are developed in parallel, and each module is developed in parallel in the participating domains (Fig. 19).

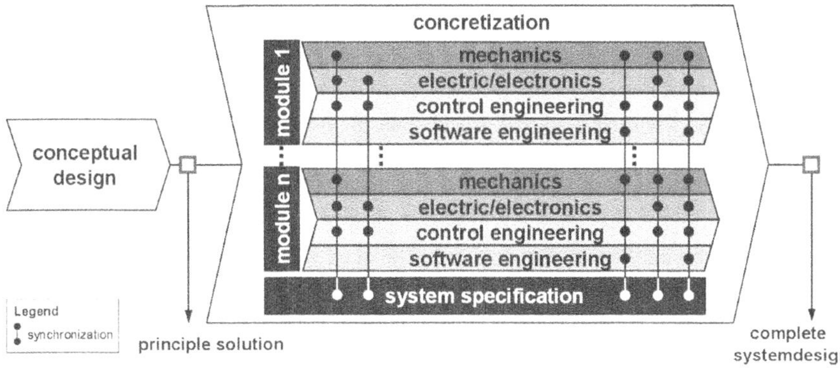

Fig. 19 Basic structure of the development process

5.1 Domain-Spanning Specification of the Principle Solution

Right from the start of our work, it became clear that a comprehensive description of the principle solution of a highly complex system needs to be divided into aspects. Those aspects are, according to Fig. 20: requirements, environment, system of objectives, application scenarios, functions, active structure, shape and behavior. The behavior consists of a whole group because there are different kinds of behavior, e.g. the logic behavior, the dynamic behavior of multi-body systems, the cooperative behavior of

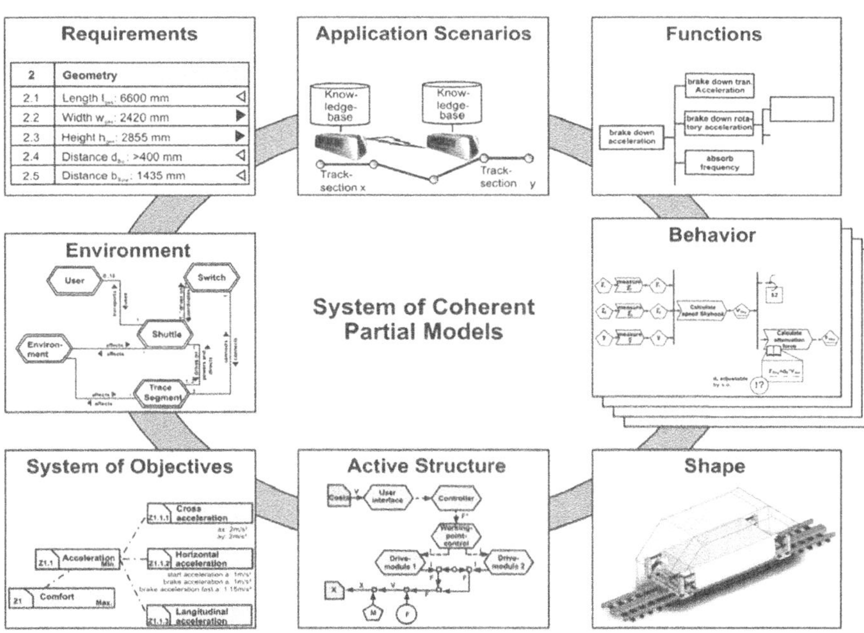

Fig. 20 Partial models for the domain-spanning description of the principle solution of self-optimizing systems

system components etc. The mentioned aspects are mapped on computer by partial models. The principle solution consists of a coherent system of partial models because the aspects are in relationship with each other and ought to form a coherent system. It is necessary to work alternately on the aspects and the according partial models although there is a certain order.

The description of the environment, the application scenarios and requirement serve as the starting point. They are usually followed by the system of objectives, the function hierarchy and the active structure. The active structure represents the core of the principle solution in conventional mechanical engineering. The modeling of states and the state transitions and also the impacts on the active structure play a decisive role in the specification of a self-optimizing system. This kind of modeling takes place within the group of behavior models. The following subchapters explain the partial models, the relationships between the partial models and the specific characteristic of the specification of self-optimizing systems.

The described partial models are not compiled in a predetermined sequence, but in a close interplay. Normally, you start with the partial models requirements, environment, application scenarios and system of objectives.

The core of the system's self-optimization is the self-optimization process and its effects on the system. As mentioned before, the process of self-optimization compares to a state transition of the system. Thereby the system is transformed from one configuration to another, and it changes its behaviour. For the entire description of the self-optimization process, the partial model active structure and two types of the partial model behavior are necessary — please remember there are quite a number of partial models representing behaviour. The two types, that are relevant here, are behavior — states and behavior — activities.

The **partial model behavior — states** describes all relevant states of the system and all events, which could cause a state transition.

The **partial model behavior — activities** specifies the cross-domain behavior of the system. The behavior is described by operation and adaptation processes. **Operation processes** characterize the activities of a system in one specific state: These are for example the acquisition of information about the environment, the derivation of adequate control interventions, and the controlling itself. **Adaptation processes** specify the activities, which are necessary for a state transition. This is the self-optimization process.

The **partial model active structure** comprises all relevant configurations of system elements (controllers, sensors, actuators) und their interrelationships. These interrelationships express flows of energy, information and — if needed — material. In each state, one definite configuration of system elements is activated. State transitions are often realized by additional system elements.

These three partial models are strongly interdependent: To each state of the system in the partial model behavior — states an operation process in the partial model behavior — activities is assigned, as well as a configuration of system elements in the partial model active structure. And to each event in the partial model behavior — states an adaptation process in the partial model behavior — activities and realizing system elements in the active structure are associated.

5.2 Managing the Development Process

Communication and cooperation between all engineers involved in the development process are essential for a successful and efficient development of advanced mechatronic systems. Within the conceptual design phase, domain-spanning development tasks and their results have to be visualized, so that developers can further elaborate the system design in a cooperative way. In the concretization phase, developers work independently from others on modules in different domains. Their specific development tasks need to be synchronized with those of other domains or modules. Therefore, domain- and module-spanning coordination processes need to be defined and communicated between developers by means of appropriate visual tools. We identified the following user tasks that need to be supported by a visual presentation tool:

Overview and details: Developers have to synchronize their work with other activities in the development process and need to identify which process steps relate to others and which can be refined and executed independently. This requires a visual overview of the complete development process as well as a focus on specific and detailed information of individual development tasks and their results.

Search, filter and results: The large number of process steps requires an efficient search mechanism. Arbitrary types of process elements should be selectable by filter operations and results should be represented and accessed efficiently.

Interactive navigation and presentation: Communication between developers is best realized by means of an interactive visualization of the complete system model. Users must be able to efficiently navigate through the complex model and present selected elements at an arbitrary level of detail. Moreover, it should be possible to present a sequence of user-defined views of the system model, for example to discuss a specific workflow.

The aim of the visual presentation tool is support the design team in managing the complex development process of advanced mechatronic systems. With such a tool, we break new grounds in process visualization by applying methods and techniques from classical design review not only to the results of individual development phases, but rather to the entire development process itself.

We chose a complex application example by selecting the complete design model of the RailCab prototype, in order to demonstrate the productivity of the visual presentation tool. The development process of the RailCab comprises of 850 development steps including about 900 development objects, e.g. kinematic models, dynamic models, controller models, state charts, list of requirements, test results etc. During the concretization phase seven modules are developed in detail. To facilitate this process, we developed a zoomable user interface that visualizes the complete development process in high detail on a high resolution projection system [13] (Fig. 21).

Figure 22 illustrates, how the visual presentation tool could look like. A zoomable view of the entire development process supports a quick navigation through the complete system model. The user can efficiently search the complete system model, navigate to any process steps of the entire process model, switch between arbitrary level of detail and present results of individual process steps, like for instance a partial model of the principle solution.

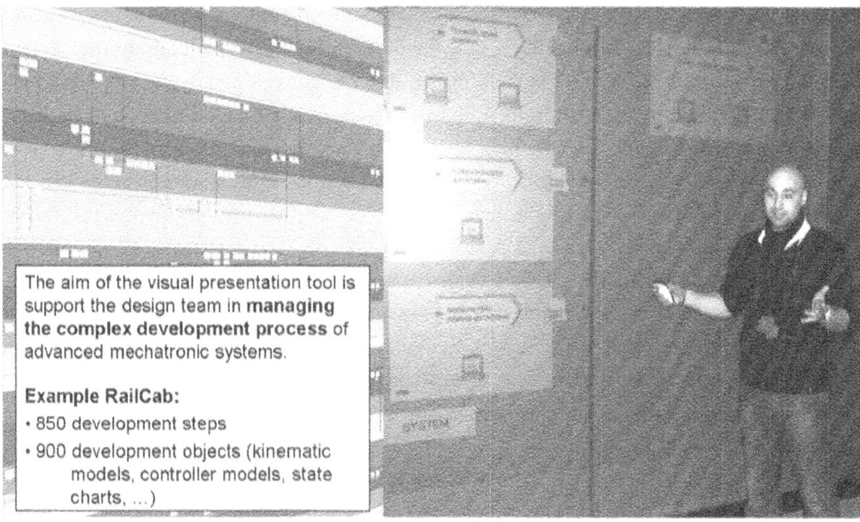

Fig. 21 User in front of a presentation wall (right) and overview of complete development process of the RailCab (left)

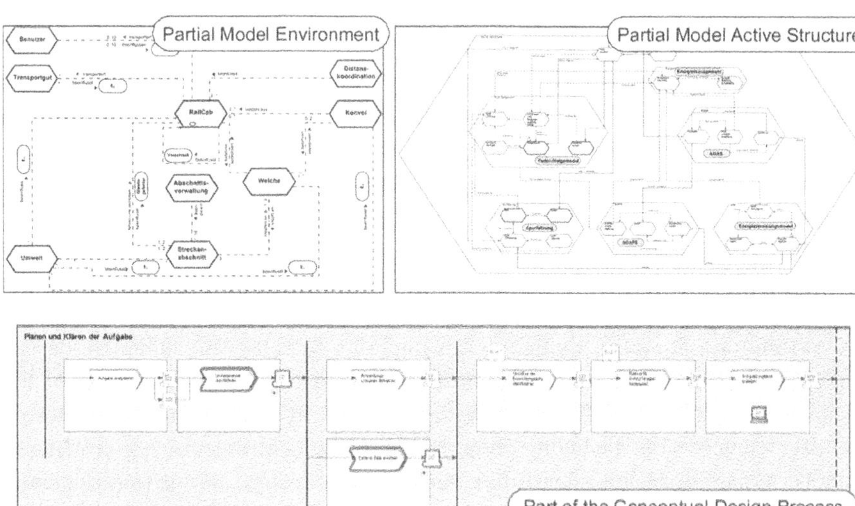

Fig. 22 Visual presentation tool for managing the development process. Partial model environment (top left), overview of the active structure (top right) and extract from the development process of the RailCab (lower half)

6 AR- and VR-based Testing of Advanced Mechatronic Systems

For the development of an advanced mechatronic system like the RailCab, we first apply the specification technique, which is introduced in the previous section, to define a principle solution of the RailCab. Based on the principle solution, we then develop numerous virtual prototypes and test beds for real prototypes, in order to analyze and evaluate the behavior of selected RailCab components and modules. However, the analysis and evaluation of the RailCab and its components can get complex and time-consuming, due to two main reasons:

- The RailCab can show a quite complex behavior, e.g. when a RailCab joins or leaves a convoy. These processes elapse fast and involve numerous variables and data, making it difficult for the engineer to comprehend the course of action and to maintain an overview of the whole procedure.
- Some prototypes of the RailCab components, e.g. the test bed for RailCab's undercarriage, operate swiftly and their parts mostly move by merely a few millimeters. For the engineer, such tiny motion is hard to perceive.

For these reasons, we apply two approaches, one AR- and one VR-based, to facilitate the analysis and evaluation of selected prototypes and their test beds:

- For the AR-based approach, we combine variables and test data from the prototype and its test bed and utilize the technology AR to augment the engineers view on the system by superimposing relevant information into the user's view.
- For the VR-based approach, we apply the technology VR to visualize missing components, like e.g. the test track, in order to simulate, how well the undercarriage prototype performs on alternative test tracks, other than our real test track on the campus in Paderborn.

In the following subsections, we describe these two examples, which demonstrate our AR- and VR-based approaches to facilitate the analysis and evaluation procedures for two prototypes of the RailCab.

6.1 AR-based Visual Analysis of the Convoy Behavior

In the first example, we apply the technology AR to facilitate the analysis of the convoy building process of several RailCabs on the test track on the university campus in Paderborn. As we only have one real RailCab prototype available for driving on the test track, we need to simulate additional virtual RailCabs to be able to simulate convoy building processes. Based on AR-technology, we project the virtual RailCabs into a video of the real RailCab which is driving over test track. The AR-application combines and visualizes all relevant information directly on screen, so that the engineer can easily observe the convoy building process on one monitor. Furthermore, the AR-application allows to check, how well an alternative version of the control software cooperates with previous versions during a convoy building process.

To evaluate the concept of convoy driving and the convoy controller under realistic conditions, a hardware-in-the-loop-system has been developed [14]. In this HIL-system, a real RailCab and four simulated RailCabs have been combined to a convoy. The principle is shown in Fig. 23. The real RailCab is driving over the test track at the university campus and serves as the convoy leader. The convoy control unit inside this vehicle manages the entire convoy. The concept of autonomously driving convoys requires the RailCabs to drive consecutively in a distance of less than one meter without any mechanical coupling [15–17]. To realize this, the RailCabs must communicate among each other. Therefore, the leading RailCab transmits its position and leading speed to the four simulated vehicles, which use this for guidance. All sensors, actuators and the mechanic parts of the virtual RailCabs are simulated. To avoid collisions between the RailCabs inside the convoy, each vehicle is distance-controlled by a subordinated speed controller. The distance is regulated relative to the position of the RailCab driving ahead.

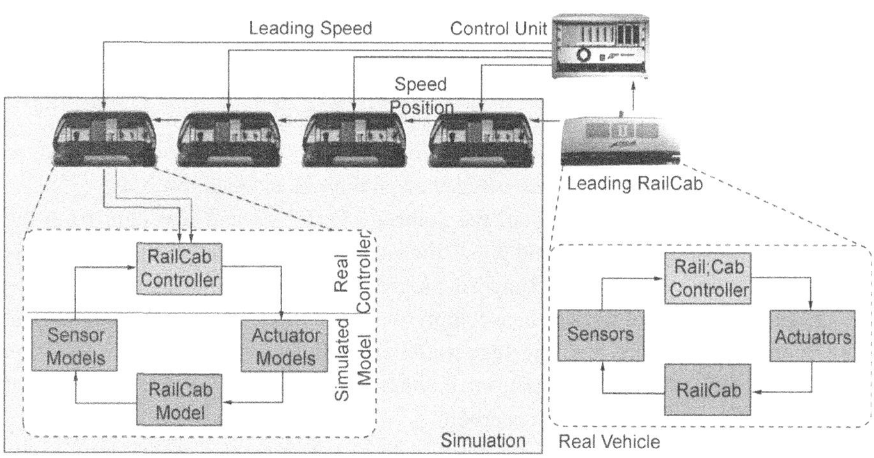

Fig. 23 Principle of the hardware-in-the-loop test of the RailCab convoy

The limitation of the HIL-simulation is that the simulated RailCabs are not visible on the test track. Their movement can only be illustrated by a timeplot. However, when using the HIL-based test bed, it is difficult to test different control strategies and parameter settings interactively during online experiments. Usually, the engineer has to conduct an experiment at the test track and collect and save the data for later analysis in the lab. Changing the strategies or parameter settings interactively during an ongoing experiment is limited and complicated. An AR-based visualization of the simulated RailCabs and corresponding data, that explains their behavior, enables an interactive testing of the HIL-simulation.

For this reason, we have developed an AR-application to visualize the simulated RailCabs and additional parameters on the real test track. We facilitate a visual analysis of the convoy behavior at the HIL-based test bed. This enables the engineers to understand the behavior of the HIL-test during an online-experiment. Main question for analysis is: Is the convoy able to drive without any collision, using different control strategies and parameters? For this, we visualize the shape of the simulated RailCabs and their motion on the real test track, as well as abstract control data such as position and speed of each RailCab, the distance between two RailCabs, braking distance and leading speed. In the following, we present four subjects for analysis, in order to illustrate how the AR-based visualization facilitates the testing of the HIL-based test bed and the evaluation of the convoy behavior.

Convoys: One important question for an energy-efficient operation of the RailCabs is, how well a convoy can be established and maintained. Figure 24 shows a convoy consisting of a real RailCab being followed by three virtual RailCabs. Seeing the simulated RailCabs superimposed on the test track help the engineers to recognize, whether the RailCabs drive in a convoy or whether a RailCab loses contact to a convoy. In Fig. 24 the last RailCab has lost contact to the convoy and the gap between the last RailCab and the preceding convoy is too large to take advantage of the slipstream.

Fig. 24 AR-visualization of a RailCab losing contact with the convoy

Collisions can occur, when RailCabs try to minimize their distance to preceding RailCabs in a convoy, in order to reduce air resistance and save energy. Figure 25 illustrates the AR-based analysis of a collision between two RailCabs. If two RailCabs collide, their shapes are highlighted in red. If the distance between two RailCabs falls below a specified limit, a collision warning indicates an upcoming possible collision. In the visualization, this is indicated by a flashing RailCab.

The **control quality** is an important indicator that helps to avoid critical conditions while driving in a convoy. In the AR-based application, we visualize such abstract quantities by plausible and easy-to-understand color-codes. Figure 26 shows a visualization of the control quality of each RailCab using such a color-code.

Here, RailCabs are highlighted in a color range spanning from green over yellow to red. The color shades indicate, how well the controller in each RailCab of the convoy is able to follow any reference values, the superordinated convoy controller in the leading RailCab, which functions as the convoy leader, communicates to all convoy members.

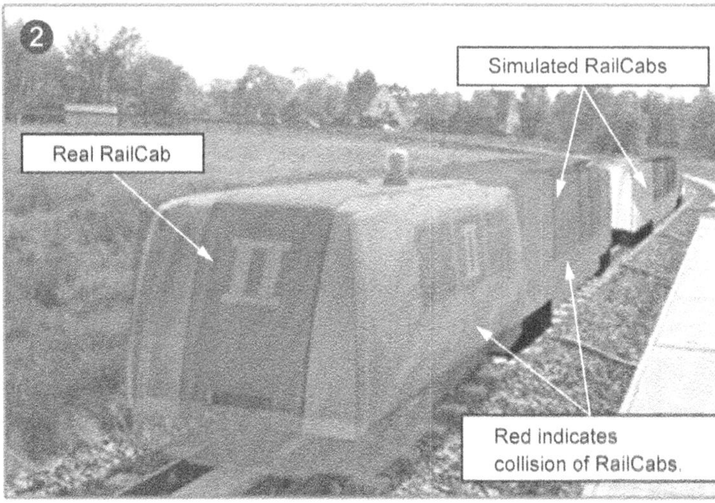

Fig. 25 AR-based analysis of a collision occurring in a convoy

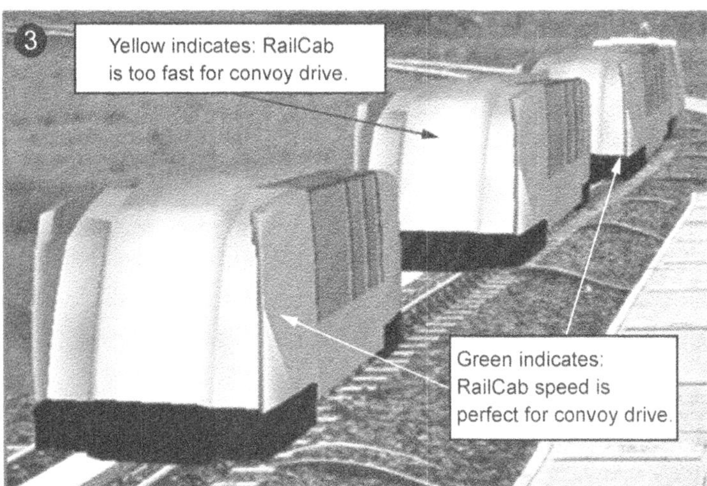

Fig. 26 Supervision of control quality

Braking distance: In case of a system or component failure, one of the crucial parameters for the convoy controller is the braking distance of each RailCab in the convoy. The braking distance directly influences the convoy behavior and helps to prevent rear-end collisions for a save operation of the convoy. For a save convoy drive, the braking distance of a following RailCab in a convoy must, at any time, be smaller than the braking distance of the preceding RailCab. To visualize the braking distance, a red bar is superimposed on the track (see Fig. 27). The highlighted track section indicates the estimated braking distance of the RailCab. The vertical red bar at the end of the highlighted track section indicates the expected final position at which the RailCab will come to a complete stop.

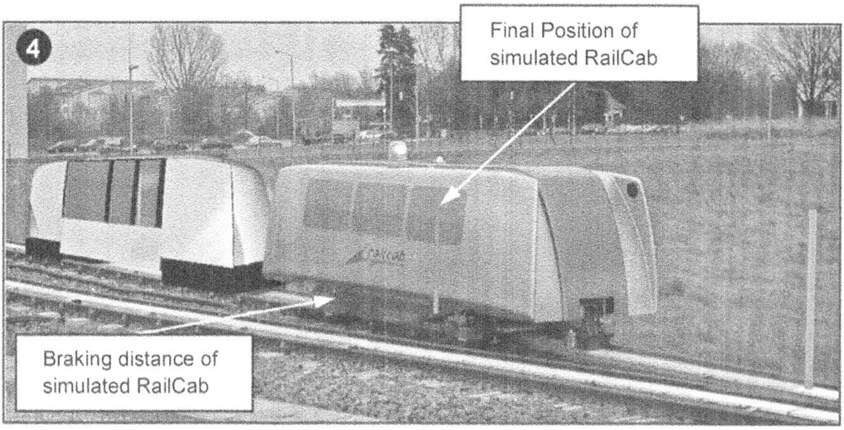

Fig. 27 AR-based visual analysis of the breaking distance and the final position of a RailCab

6.2 Realization of the Augmented Reality System

Figure 28 shows the HIL-test bed setup on the left and the setup of the AR-system on the right. A communication server is integrated between both components. The lower half of Fig. 28 shows the operator's control desk for observing the test track.

The HIL-test bed is composed of a real RailCab and a real-time prototyping system, produced by dSpace. A simulation of four real RailCabs is running on the real-time system under hard real-time conditions. The real RailCab on the test track is the leader of the simulated convoy. This RailCab communicates with the simulation system by an internal, proprietary bus system.

The simulation system sends position and speed data of all RailCabs to the communication server. The server is used as an interface between the HIL-test bed, running under hard real-time conditions, and the AR-system, running under soft real-time conditions. Every 10 ms the HIL-test bed sends data of the RailCabs to the communication server. Since the AR-system just renders a new image every 40 ms, it gets each fourth data package from the communication server.

The AR-system is built up from a PC with a Pentium 4, 2.8 GHz Dual Core processor, a NVIDIA 8800 GTX graphics board and 2 GB RAM. The output device for the video is a HDTV monitor with a wide screen ratio of 16:9. For recording the video of the test tack, a Sony HVR-V1E HDV video camera is used. The monitor and the camera both have a resolution of 1980 × 1080 pixels. We need the high resolution to be able to monitor not only RailCabs that pass by near the camera, but also those RailCabs that drive on the far side of the test track.

The AR-software application was implemented in C++ and the 3D graphics toolkit OpenSceneGraph, which offers various functions to render animated 3D objects. The video image is rendered as a 2D image in the background with the fragment shader of the graphics board. To demonstrate the simulated RailCabs in the AR-application, CAD models of the real RailCab were integrated into the application.

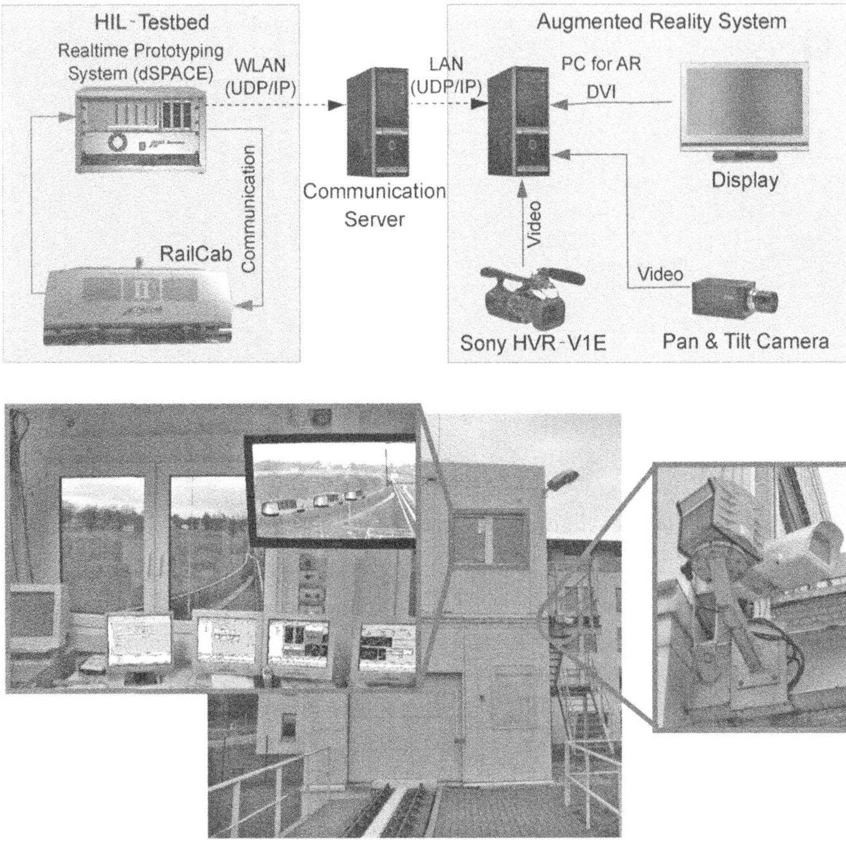

Fig. 28 Setup of the augmented reality system at the test track

6.3 VR-based Visual Analysis of the RailCab Undercarriage

The second example describes the use of the technology VR, in order to facilitate a targeted testing of the RailCab undercarriage components as well as for the controller development. A test bed for the RailCab undercarriage in scale 1:2.5 has been developed at University of Paderborn. The RailCab undercarriage actively steers over passive switches, enabling the RailCab to choose its direction at a switch.

As the wheels are not forced along the tracks, like with conventional trains, guidance along a track and steering over a switch require frictional forces. These are particularly important on wet or icy tracks. This led us to developing a camber-variable undercarriage, which actively controls the camber of the wheels. The aim is to uncouple guidance and steering forces from the frictional forces between wheel and track. In addition, the undercarriage actively steers each wheel individually, in order to reduce torque and frictional forces considerably when steering the RailCab over a switch.

The principle is shown in Fig. 29. On the left, Figure 29 shows a profile of a track and a wheel with varying camber angles and without changed camber angle. The track has a spherical cross cut profile, so that the wheels can support themselves on it while cambering.

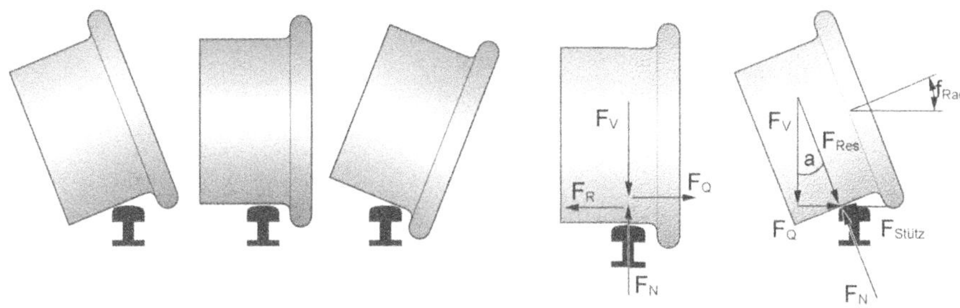

Fig. 29 Varying camber angles of a wheel (left) and distribution of forces at a wheel in normal position and an inclined wheel (right)

On the right of Fig. 29 it is shown, how the forces are distributed at a wheel in normal position and an inclined wheel. If the wheel is standing up straight, the RailCab is kept by the friction force F_R on the track. The friction force compensates the shear force F_Q, which can e.g. be generated by centrifugal force. When a wheel cambers, it supports itself on the track mechanically. The shear force is compensated by the profile lateral force $F_{Stütz}$ at the contact point of the wheel, but not by the friction anymore.

6.4 Test Bed of the Undercarriage

Figure 30 shows a RailCab undercarriage and its test bed. The test bed of the undercarriage comprises a wheel suspension of the RailCab, which is fixed in a frame. In the center, two actuators, which adjust the camber angle of the wheels on both sides of the undercarriage, are mounted to the frame. Each wheel is connected to another actuator that adjusts the wheel's steering angle. The wheels are positioned on a short piece of a track. Sensors for measuring the distance between wheel and track are mounted to the undercarriage.

For the active steering of the wheels of the undercarriage, a controller computes the camber and steering angles of each wheel in real time and operates the actuators for setting camber and steering angles as well as distance sensors for measuring the position of the wheels on the track.

Since the physical test track is not directly visible from the test bed, it is not possible to simply map the dynamic behavior of the undercarriage prototype on the test bed with the RailCab prototype on the test track. This handicaps a targeted analysis of the prototype's dynamic behavior at the test bed, since data first needs to be collected and evaluated after tests have been finished. In general, the prototype's dynamic behavior can however be evaluated without any specific visualization. But since there is no representation of the test track, which facilitates mapping steering and inclination motion at the test bed to the course of the test track, evaluation is clearly more difficult.

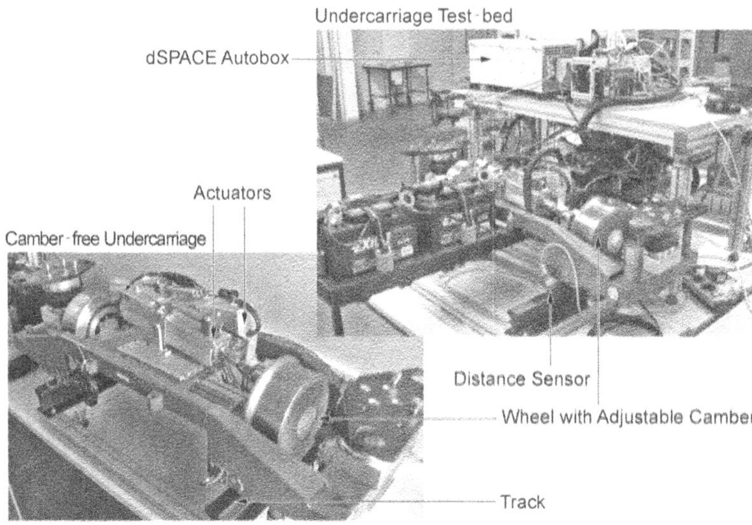

Fig. 30 Camber-variable undercarriage and test bed

6.5 Virtual Reality Application for Visualization

We have developed a virtual reality-based application to facilitate a fast allocation of the camber-variable undercarriage behavior to the course of the test track. The VR-application comprises a visualization of a virtual test track and a virtual model of the camber-variable undercarriage. The application is connected with the undercarriage test bed via a real-time bus. This ensures that steering and inclination angles of the wheels in the test bed correspond with the virtual model of the undercarriage in the VR application. Figure 31 provides an overview of the entire system. The virtual test track serves as a track model for the undercarriage test bed, providing the course of the track, including all slopes, courses of curves, switches, etc., the undercarriage is driving over. The undercarriage adapts the steering and camber angles of its wheels in the test bed according to the course of the virtual test track. While driving along the virtual test track, the VR model of the undercarriage also adjusts the steering and camber angles accordingly and thus demonstrates the dynamic behavior of the prototype in the test bed, as if this would be driving along the virtual test track.

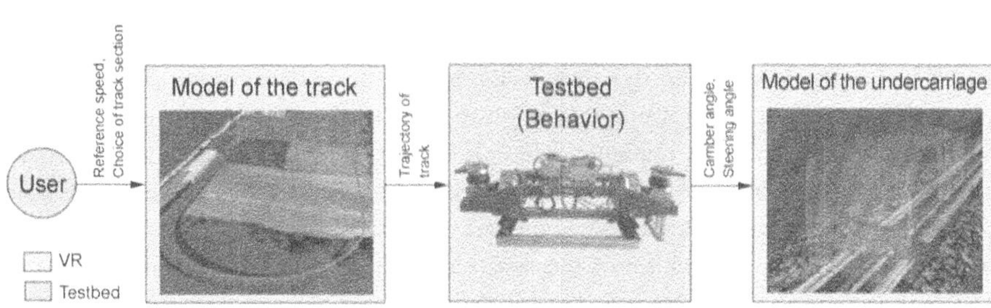

Fig. 31 Overview of the VR application for visualizing the undercarriage on a virtual test track

The VR environment comprises a virtual model of the RailCab test track, which is located on campus at University of Paderborn [15, 18]. Figure 32 right shows on overview of the test track. The track consists of 2 loops, which are connected with each other by 2 switches. The right hand loop has a length of approx. 430 m with a maximum slope of 4,3 %. The left hand loop is about 440 m long with a maximum slope of 3,0 %. While the left hand loop exists only as a virtual model, the right hand loop has a real counterpart on campus at University of Paderborn.

Fig. 32 3D-model of camber-variable undercarriage (left) and virtual test track (right)

On the left, Figure 32 shows the 3D-model of a RailCab with its camber-variable undercarriage on the virtual test track. The RailCab is 3 m long, 1,2 m tall and 1,5 m wide. It is rendered transparently, in order to reveal the camber-variable undercarriage, which is located below the RailCab chassis. The 3d-model of the undercarriage was extracted from 3D-CAD data. All major parts that are essential for demonstrating the functional behavior, such as wheels, actuators for steering and camber angle adjustment, frame, single wheel suspension and sensors, were integrated into the visualization.

For rendering the VR-environment (cp. Fig. 32, right) we use the C++-based scenegraph API OpenSceneGraph, which provides basic elements for creating and efficiently administrating graphical objects for a 3D visualization.

Within the test bed, the undercarriage is activated by data from a simulated test track. The resulting response of the undercarriage can be studied directly at the test bed. As the test track is just simulated and not visible at the test bed, the engineer cannot directly correlate the course of the test track, being the source of the actuation, with the response of the undercarriage, i.e. its' steering and camber angles.

For this reason, we developed a VR-application, which shows a 3D-model of the undercarriage on the virtual test track, in order to facilitate a fast correlation of the dynamic behavior of the camber-variable undercarriage and the corresponding section of the test track the undercarriage is passing over. The VR-application is connected to the undercarriage test bed and visualizes the steering and camber angles of the wheels based on the current course of the track. This allows to cross-check the motion of the undercarriage relative to the virtual test track and to analyze critical driving situations, e.g. when the RailCab is driving over a switch.

Figure 33 shows an overview of the system. The VR-based visualization is connected with the real time controller of the test bed. That way, the visualized camber and steering angles of the virtual undercarriage correspond to the test bed. The virtual test track provides data of the course of the track (e.g. inclinations, drops, curvature, switches, etc.) based on which the real time controller of the undercarriage computes corresponding camber and steering angels for each wheel. These angles are sent to the actuators in the undercarriage test bed as well as fed back into the VR-simulation for direction comparison. That way, the simulated drive over the virtual test track can be analyzed on the undercarriage test bed.

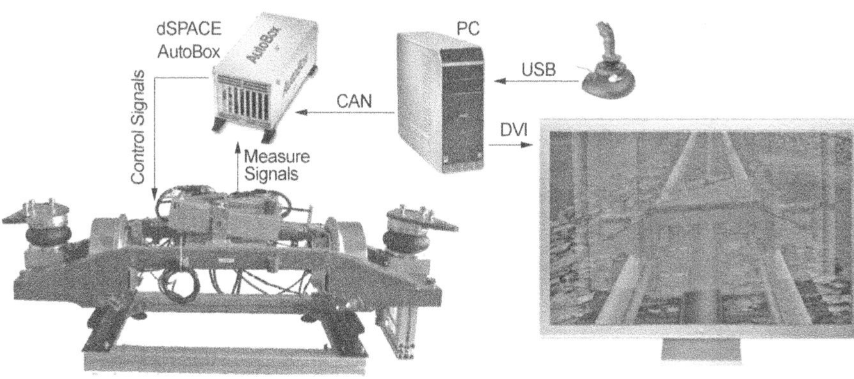

Fig. 33 Overview of the VR-based system for testing the undercarriage on different virtual test tracks

The VR-based testing application provides an interactive steering mode, where the engineer can control the steering angles via a joystick and study the effect of the steering command on the resulting camber angles and the overall behavior of the undercarriage. Also, the speed or the desired driving direction at a switch can be controlled, in order to study the system response.

6.6 Visual Analysis of the Behavior

We have conducted numerous test drives, for which the undercarriage test bed was connected to the VR application for visualizing the simulated test drives along a virtual test track. We performed visual analyses of the test drives, where we focused on examining steering and camber angles and analyzing the dynamical behavior of the undercarriage on the test track and especially during the process of crossing a switch.

Visualization of steering and camber angles: Within the VR-application, we visualize the camber and steering angles of each wheel, in order to facilitate a better understanding of the behavior of the camber-variable undercarriage. Figure 34 shows two screenshots from the VR-application with a visualization of the camber angles of the wheels, as provided by the real time controller and being set by the actuators. The VR-application shows the undercarriage in a close-up view. That way, during a simulated drive over the test track, the resulting motion of all moving parts within the undercarriage can easily be analyzed by the engineer. Hardly visible objects, like for instance small moving parts or abstract information, are visualized by plausible illustrations.

On the left of Fig. 34, camber angles values are visualized by the green segment within the yellow semicircle which indicates the maximum valid range for the camber angles. Analog to this, steering angles are shown on the right of Fig. 34. Here, the same visualization metaphor is used to illustrate the steering angles with a different maximum valid range and different colors.

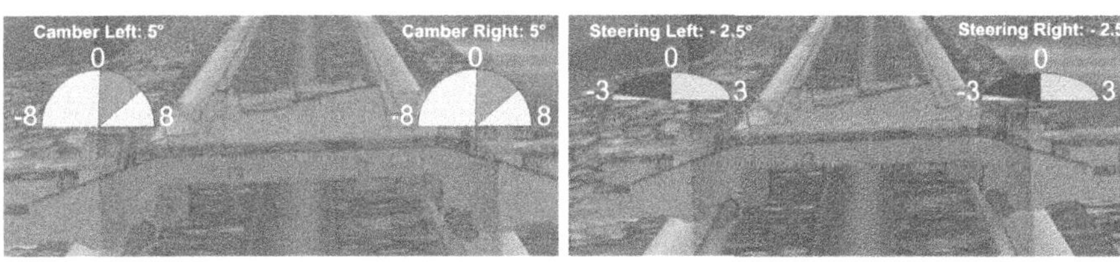

Fig. 34 Visualization of camber (left) and steering angles (right) of two wheels in the RailCab undercarriage

Analysis of the undercarriage's dynamical behavior: A fast information acquisition at the physical test bed is crucial for this work. This applies particularly for mechatronical systems that operate under hard real-time conditions and which adapt their behavior within a split second. The aim of the visualization is, to demonstrate the camber-adjusting behavior of the undercarriage during a simulated drive. Since there is no test bed available directly at the test track, it is difficult to identify a correlation between the driving behavior of the undercarriage and the test track.

We describe the analysis of the undercarriage's dynamic behavior in the following example. In this example we want to analyze the algorithms and strategies for driving over a switch. Main question here is, when does the RailCab start to adjust the steering and camber angles of its wheels and when does it turn the wheels back into normal steering position, as well as back to neutral camber angles? This is an important question, since considerable time elapses from the moment, the controller passes the set points to the actuators until the actuators have accomplished adjusting the steering and camber angles in the undercarriage. This needs to be optimized by adjusting the controller parameters.

Analysis of a switch-crossing: Figure 35 shows a RailCab, which is crossing a switch. In Fig. 35 top left, the RailCab is approaching the switch, while the steering angle is still 0°. When the RailCab drives into the switch a steering angle is calculated, which is then set in the test bed setup of the undercarriage and is also visualized in the VR-application. Due to this, the wheels turn to the left (cp. Fig. 35 top right). As the RailCab enters the switch, the course of the track turns from slight left to slight right. The wheels need to steer along this change of direction and also turn slightly from left to right (cp. Fig. 35 top right and lower left). When the RailCab exits the switch, the tracks turn straight again and the steering angle returns to 0° (see Fig. 35 lower right).

While we develope the control algorithms for the actuators, we still need to optimize the dynamic behavior of the undercarriage. For instance, it may well be possible that, while the RailCab enters a switch, a steering wheel is determined that turns the wheels to the left. Since the RailCab keeps on driving during the process of setting the steering angle, the steering angle is not completely adjusted until the wheels already need to turn back to the right again. At the test bed, the visualization of the test track helps to easily identify and solve this problem. This facilitates a targeted analysis of control algorithms and strategies.

Before passing a switch, steering angle =0.0°

Passing over a switch, steering angle = −1.28°

Passing over a switch, steering angle =1.57°

After passing a switch, steering angle =0.0°

Fig. 35 Analysis of a switch-crossing: Approaching the switch with steering angle 0° (top left), entering the switch with steering angle −1.28° (top right), crossing the switch with steering angle 1.57° (lower left) and exiting the switch with steering angle returning back to 0° (lower right)

7 Conclusion

The development of advanced mechatronic systems is a challenging task which requires know-how of numerous experts from various disciplines. Our new specification technique for the design of such systems facilitates this interdisciplinary process by combining the contribution of each discipline and applying one common specification technique for the conceptual design. We enhance the development of advanced mechatronic systems by integrating methods and techniques from classic design review into a visual presentation tool and applying these methods to the development process itself.

Based on the experience we gained, when applying the AR- and VR-based approaches to some of RailCabs' prototypes and test beds, we can confirm that these approaches facilitate a fast visual analysis and a targeted testing of future advanced mechatronic systems. However, integrating AR- and VR-based approaches into RailCabs' prototypes and test beds is still a time-consuming and complex task, because various simulation models and their interfaces need to be adapted. Thus, our future work focuses on automating the generation of appropriate AR- and VR-based visualizations as well as automatically connecting the visualization components with corresponding simulation models of the prototypes and test beds.

Acknowledgments

This contribution was developed and published in the course of the Collaborative Research Center 614 "Self-Optimizing Concepts and Structures in Mechanical Engineering" funded by the German Research Foundation (DFG) under grant number SFB 614.

References

[1] Gausemeier J, Frank U, Steffen D (2006) Specifying the principle solution of tomorrow's mechanical engineering products. In: Proceedings of the 9th International Design Conference DESIGN 2006. May 15–18, Dubrovnik, Croatia

[2] Gausemeier J, Ebbesmeyer P, Eckes R (2006) Virtual Production — Computer Model-based Planning and Analyzing of Manufacturing Systems. In: Dashchenko, A.I. (ed): Reconfigurable Manufacturing Systems and Transformable Factories, Springer

[3] Shen Q, Grafe M, Bauch J, Radkowski R (2005) Interdisciplinary knowledge sharing in solution elements based virtual prototyping of mechatronic systems. In: Proceedings of the 9th IEEE International conference on CSCW in design, IEEE Press, Coventry, UK

[4] Shen Q (2006) A Method for Composing Virtual Prototypes of Mechatronic Systems in Virtual Environments. Dissertation: University of Paderborn, Heinz

Nixdorf Institute, HNI-Press, Paderborn

[5] Ibrahim F, Schmidt C, Klein M (2005) Predictive Advanced Front Lighting System, In: Proceedings of the ISAL 2005 Symposium, Darmstadt University of Technology, Darmstadt, Germany

[6] Berssenbruegge J, Bauch J, Gausemeier J (2006) A Virtual Reality-based Night Drive Simulator for the Evaluation of a Predictive Advanced Front Lighting System. In: Proceedings of the ASME 2006, Philadelphia, Pennsylvania, USA

[7] Berssenbruegge J (2006) Virtual Night Drive — A method for Displaying the Complex Light Distribution Characteristics of Modern Headlight Systems Within a Simulated Night Drive. Dissertation: University of Paderborn, Heinz Nixdorf Institute, HNI Press, Paderborn

[8] Gausemeier J, Radkowski R, Oehlschlaeger H, Krebs J (2005) Development of a Mobile Augmented Reality Test Platform for the Analysis and Evaluation of Automotive Ergonomics. In: Proceedings of ISMAR

[9] Lückel J, Hestermeyer T, Liu-Henke X (2001) Generalization of the Cascade Principle in View of a Structured Form of Mechatronic Systems. IEEE/ASME AIM 2001, Villa Olmo, Como, Italy

[10] Frank U, Giese H, Klein F, Oberschelp O, Schmidt A, Schulz B, Vöcking H, Witting K, Gausemeier J (ed) (2004) Selbstoptimierende Systeme des Maschinenbaus — Definitionen und Konzepte. HNI-Verlagsschriftenreihe, Vol. 155, Paderborn

[11] Pahl G, Beitz W, Feldhusen J, Grote KH (2007) Engineering Design — A Systematic Approach, 3rd edition. Springer, London

[12] Verein Deutscher Ingenieure (VDI) (2004) VDI-guideline 2206 — Design methodology for mechatronic systems. Beuth-Verlag, Berlin

[13] Geiger C, Reckter H, Dumitrescu R, Kahl S, Berssenbrügge J (2009) A Zoomable User Interface for Presenting Hierarchical Diagrams on Large Screens. In: Proceedings of the 13th HCII, 19–24 July, San Diego

[14] Radkowski R, Henke C (2007) Einsatz von Virtual Reality zur Untersuchung realer und simulierter mechatronischer Systeme am Beispiel der neuen Bahntechnik Paderborn. In: Proceedings of 10. IFF-Wissenschaftstage, Magdeburg

[15] Trächtler A (2006) Railcab — mit innovativer Mechatronik zum Schienenverkehrssystem der Zukunft. In: VDE (ed) VDE Congress, Aachen

[16] Rustemeier C, Nachtigal V (2003) RailCab — A New Modular Rail System. 4. International Conference "Trans-Mech-Art-Chem", Moskau, Russia

[17] Henke C, Fröhleke N, Böcker J (2006) Advanced Convoy Control Strategy for Autonomously Driven Railway Vehicles. In: Proceedings of the 9th IEEE Conference on Intelligent Transportation Systems, Toronto, Canada

Authors

Prof. Dr.-Ing. Jürgen Gausemeier is Professor for Product Engineering at the Heinz

Nixdorf Institute of the University of Paderborn. He received his PhD from the Institute for Machine Tools and Production Engineering at the Technical University of Berlin. His doctoral supervisor was Professor Spur. During his 12 year career in industry, Dr. Gausemeier was Head of R&D of CAD/CAM systems. His last position was Division Manager for process control systems at a well-known Swiss company. Beyond his university activities, he is active as a Member of the Board and Managing Director of the scientific society "Berliner Kreis — Scientific Society for Product Development e.V.". Furthermore, he is the initiator and Chairman of the Supervisory Board of the consultant company UNITY AG. Jürgen Gausemeier is Member of the Executive Board of "acatech — German Academy of Science and Engineering". In 2009 he was appointed to the German Council of Science and Humanities.

Dr.-Ing. Jan Berssenbrügge is a post-doc at the Department of Product Engineering of the Heinz Nixdorf Institute at University of Paderborn, Germany. He received a MSc in Computer Science in 1998 and a PhD in Mechanical Engineering in 2005 both from University of Paderborn. His research focus is on virtual prototyping and applications of Virtual Reality and Augmented Reality within the product development process.

Dipl.-Ing. Michael Grafe is Chief Engineer at the Heinz Nixdorf Institute, University of Paderborn. Since 1995, he has been building up the research area "Virtual and Augmented Reality" at the Department of Product Engineering, where he coordinates the VR and AR research activities at the Heinz Nixdorf Institute.

Dipl.-Wirt.-Ing. Sascha Kahl is, since 2007, a research assistant at the Department of Product Engineering of Professor Gausemeier at the Heinz Nixdorf Institute of the University of Paderborn. His research interest is in methodologies for the design of self-optimizing systems. His work focuses on the conceptual design phase and techniques for the specification of the principle solution.

Dipl.-Inform. Helene Wassmann studied computer science at University of Paderborn with focus on computer graphics and digital image processing. Since 2007, she is a research assistant at the Department of Product Engineering at the Heinz Nixdorf Institute. Her research focuses on visualization of technical systems.

Corresponding author

Jürgen Gausemeier (✉)
Heinz Nixdorf Institute, Fürstenallee 11, 33102 Paderborn, Germany
E-mail: Juergen.Gausemeier@hni.upb.de

From Space to the Forest and to Construction Sites: Virtual Testbeds Pave the Way for New Technologies

Jürgen Roßmann
Institute of Man-Machine-Interaction, RWTH Aachen University

Abstract

In recent years, virtual reality has emerged as a key technology for improving and streamlining design, manufacturing and training processes. Based on experience in the fields of space robotics, industrial manufacturing and multi-physics virtual prototyping, "virtual testbeds" are currently being designed and implemented. Building on experiences gained in space robotics applications, the idea of virtual testbeds currently conquers new fields of applications in the manufacturing industry, on construction sites and even "in the woods". Interestingly, all fields can be supported with a rather generic and common basis so that the different applications can be realized very cost-efficiently.

Keywords

Projective Virtual Reality, Space- and Industrial Robotics, Virtual Factory, Working Machine Simulation, Virtual Forest

1 Introduction

Simulators play a great role in space applications. In order to make space missions as safe and as successful as possible in an everything but friendly environment, simulation technology has developed significant potential for the planning, training und supervision of space missions. Whereas the required technology used to be expensive and maintenance intensive, the new PC-based systems that are being developed today, make this technology available in an economically sensible manner for the training of working machine operators.

The algorithms that bring virtual harvesters, forwarders und logging trucks to live

Fig. 1 Virtual worlds related to space applications

are the same that are being used for space simulation and today provide a realism of the simulated forest and the working machines that was not considered possible a few years ago. The background of such algorithms [1] and how robotic technology influences the development not only of simulation systems but also of new control and driver assistant systems will be a key issue in this paper.

As modern forest or construction machinery is complex, expensive and rather difficult to handle, the larger manufacturers today provide virtual training environments to familiarize new drivers with the new machines without risking the "hardware" . Virtual Reality based simulators allow not only for safe training with the virtual machines, but also support the practicing of "non-nominal" and contingency situations and thus make ideal "testbeds" for the development and training for working machines. In the new generation of such simulators, besides the visualization and operating aspects (ergonomic screen setup, original seat, board computer and operating elements) the simulation of the physical properties of the trees and the machines as well as the simulation of the interaction between the machine and the terrain are key issues for the "feeling" the driver gets from a virtual compared to the physical machine. Deriving

Fig. 2 Advanced physics simulation opens up new application fields

from space simulation techniques, the simulation of these physical properties is quite well understood today and generalized methods have been developed to simulate not only forest machines but also construction and agricultural machines. The ideas behind these algorithms originate from developments and tests in space robotics environments.

2 Simulator Applications in Space

On the way to the successful planning, deployment and operation of the International Space Station, 3D simulation and virtual testbeds are heavily being used. Under contract of the German Space Agency DLR, it has become RIF's task to provide a Projective Virtual Reality System to provide a virtual testbed built after the planned layout of the COLUMBUS module enabling astronauts and scientists to practice operational procedures and the handling of experiments. The possibility for distributed multi-user access to the virtual lab and the visualization of real-world experiment data comprise the key features of the system. Through the ability to share the virtual world, cooperative operations can be practiced easily and trainers and trainees can work together more effectively in the shared virtual environment. The ability to visualize real-world data will be used to introduce measured experimental data into the virtual world on-line in order to allow realistic interaction with the science reference model hardware: The users'actions in the virtual world are translated into corresponding changes of the inputs of the science reference model hardware; the measured data is then in

Fig. 3 The virtual world of the Columbus module, the European contribution to the International Space Station

turn fed back into the virtual world. In order to provide astronauts and scientists with an even more complete insight into various aspects of COLUMBUS and the ISS, the simulation of the COLUMBUS module is currently being extended to include the entire International Space Station.

Simulation is a vital component in the operation of most space missions. This is particularly true in the case of the International Space Station (ISS), where simulations are used in applications ranging from the training of astronauts to the evaluation of docking procedures to the verification of tasks for the station's manipulators [1–3]. Most such systems focus on the simulation of individual ISS components. The Intelligent Virtual Station (IVS) under development at the NASA Ames Research Center [4] targets the management of large amounts of information to streamline the development cycle for ISS components and has the potential to be expanded to enhance existing training and operations tools.

The simulation system CIROS® (www.ciros-engineering.de), developed by the Dortmund Center for Computer Integrated Manufacturing (RIF) in cooperation with Partners, provides a comprehensive set of functionality for simulating robot mechanisms such as kinematic chains or end-effector trajectories. With its foundations in robot simulation, CIROS® is also well suited to simulating the wide range of systems found on the International Space Station. The simulation can also be connected to external devices to provide the operator with, for example, force feedback based on the results of their actions.

Virtual Reality (VR) is a powerful tool which can be used to construct a complex and immersive yet intuitive visualisation of a simulated system. Such a visualisation can be presented with a 3D head-mounted display, projection screen or a 360°-surround CAVEtm configuration. Combining VR with other CIROS® components, such as the robotic action planning system and multi-agent control system results in the method known as "Projective VR" [2], which was applied successfully in 1999 to command the German ETS-VII Experiment, a collaborative effort between the Japanese Space Agency (NASDA), the RIF team and the German Space Agency (DLR). This method can be applied to the ISS as a training tool or for the teleoperative control of Station-based robots.

2.1 The Simulation of the International Space Station

Building on previous work on the simulation of the COLUMBUS module, the European contribution to the International Space Station, a VR model of the ISS has been developed for CIROS®, shown in Fig. 4. Modern 3D rendering techniques such as texturing and bump-mapping are used to improve the visual realism of the simulation. Using input devices such as a 3D connection's SpaceMouse, the operator is able to fly freely around the outside ISS or enter through the Joint Airlock to explore the interior.

We use the North American Aerospace Defense Command (NORAD)'s Simplified General Perturbation (SGP4) orbital prediction algorithm and two-line elements to predict the position of the ISS in orbit. Due to the computational intensity of the SGP4 algorithm, it is run once per second with the exact position interpolated between the calculated values at each frame of the simulation allowing both smooth and accurate real-time determination of the Station's position. The position of the sun with respect to the ISS is also approximated with the lighting adjusted appropriately.

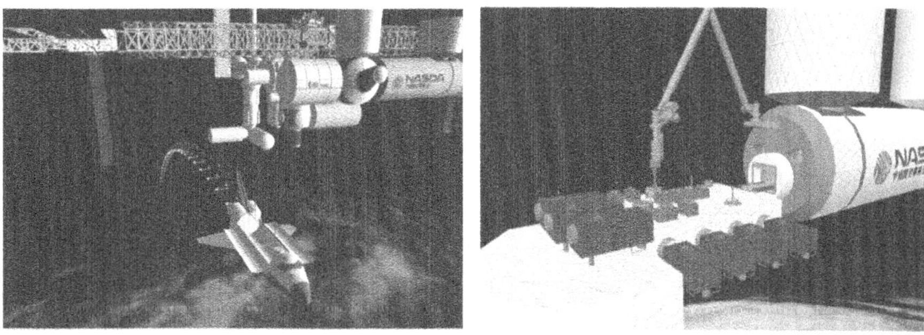

Fig. 4 The virtual world of the International Space Station

The Space Station can be explored at any stage in its assembly, from the initial Zarya Control Module through to completion, with the operator being able to select the desired configuration by date or mission. Alternatively, an automatic assembly mode can be activated to provide a walkthrough of the entire construction process.

Fig. 5 The International Space Station at the maximum level of detail

Taking advantage of the virtual nature of the simulation, it is possible to provide the operator with additional audiovisual information to augment the VR model. A "heads-up display" (HUD) can be activated to provide the user with information about a specific assembly stage or component of the Station. The speech generation capabilities of CIROS® can be used to give a verbal introduction to an ISS module as it is entered. In a training situation, remaining tasks can be displayed on the HUD while further audio instructions are provided either automatically via the speech generation server or directly by an instructor with a microphone.

2.2 The Virtual Human

The very presence of a humanoid component in the simulation provides a familiar frame of reference. The ability to realistically simulate anthropomorphic motions provides several additional advantages.

The simulation and control of an anthropomorphic kinematic system, or "Virtual Human" , is a daunting challenge. It too can be approached from a robotics perspective by regarding the kinematic system as a multi-agent system consisting of several coupled kinematic chains (i.e. the arms, legs, head and body). The Virtual Human has been modelled in this manner for CIROS® using the concept of the IRCS as a control system.

In a training scenario, the Virtual Human can take on a variety of roles. As an instructor, it can provide useful suggestions or criticism of the user's actions by following a predetermined set of rules and triggers, comparing the user's actions to its own optimised action plan or acting as an avatar for a real instructor. It can also be commanded to perform a task for the trainee, demonstrating the correct order of actions or introducing a new concept. The Virtual Human can alternatively act as a teammate, assisting the trainee in completing their tasks.

Another application of the Virtual Human is in evaluation of ISS workstations, as in Fig. 6. The complexity and feasibility of a task can be determined by commanding a Virtual Human to perform the task in simulation. The resultant actions can be used to identify possible timesaving optimisations or inefficiencies in the layout of a module. Despite the weightless environment of the ISS, ergonometric considerations are important for astronauts working in space for extended durations. The ergonomy of a workstation can be investigated by calculating the load and strain on, for example, the Virtual Human's spine as it executes tasks. The effects of phenomena such as the loss of muscle and bone mass in astronauts can be simulated in this way by changing the parameters of the Virtual Human's kinematics over time.

Fig. 6 The Virtual Human inside the International Space Station

2.3 The Concept of Anthropomorphic Multi-agent-Systems

The ideas originally developed for ergonomic simulations in space can also be transferred to industrial application, if the human kinematics are dealt with as multi-agent systems. The central idea of anthropomorphic multi-agent-systems is the human-like coupling of individual kinematic chains, which represent the limbs and the trunk of the human body (see Fig. 7). The implementation of the individual kinematic chains builds on methods and algorithms well-known in common robotics. Thus, modeling, programming and simulating the kinematic chains can be managed with the same set of tools used for industrial robots.

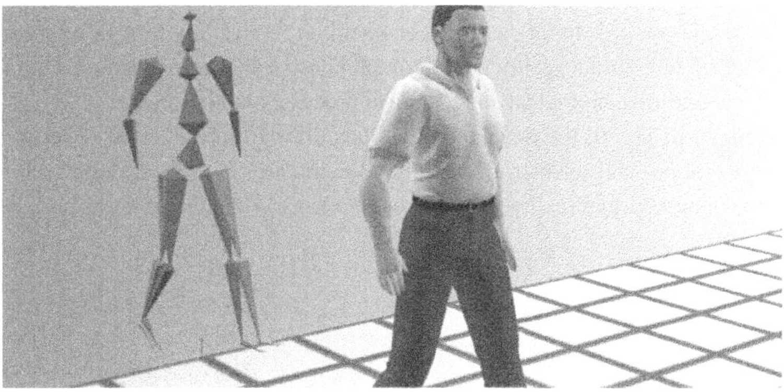

Fig. 7 The Virtual Human — modeled as an anthropomorphic robot

Although, beyond the scope of robots in industrial applications, the individual kinematic chains in this approach are controlled as multi-agent-systems in order to meet the requirements of sophisticated human-like motions. Each of the individual kinematic chains is guided by a control which exhibits the characteristic features of an agent. The agents accomplish given tasks independently by evaluating models of their environment and the kinematic chains assigned to them. In order to provide motion control for the kinematic chains, the agents are not addressing the anthropomorphic properties of their configuration as a multi-agent-system, since they only consider parameters of kinematic chains, which are also used to describe the range of properties of common industrial robots. These parameters and descriptions are chosen to make use of methods widely known in robotics.

Fig. 8 The Virtual Human allows to evaluate workplaces from an ergonomic point of view

3 The Virtual Forest

Apart from well-know fields of application in space and industry, new fields are currently being tackled. The Virtual Forest is a joint effort of various partners who combine their know-how in the fields of automation/robotics, machine development and forestry to realize a virtual reality based comprehensive framework in order to identify, visualize and optimize biological and technical processes in the forest.

As shown in Fig. 9, the developed methods draw from the fields of aerial survey technology, space- and terrestrial robotics, localization and navigation technology, sensor technology, virtual reality and - of course - latest silvicultural know how.

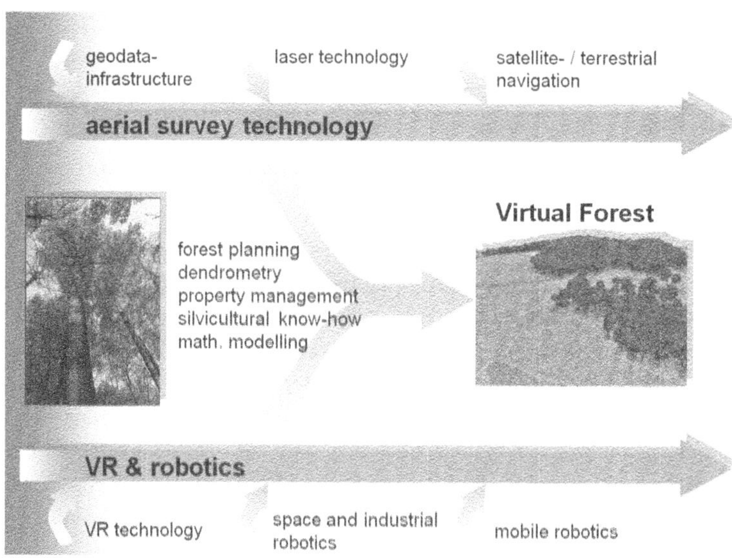

Fig. 9 The Virtual Forest as a new VR-based approach to forestry

The overall goal of the virtual forest is to develop a comprehensive database with a versatile structure which is able to hold all forest relevant data with a very high level of actuality. This database has to be filled in several steps to provide the necessary basic knowledge about the forest. The general idea is to use latest remote sensing technology, i.e. to fly an airplane or a drone to gather photogrammetric and laser based measurements about a forest. The sensed data then undergoes various image processing, object identification and classification procedures in order to derive tree-wise as well as stand-wise properties of the measured wood resources. The key to the success of this step is that a "multi-sensor-approach" is used which combines the data processing algorithms which have been and are being developed in cooperation with forest experts from all over Germany. These are currently being integrated into the Virtual Forest framework whose key design issue is to allow the algorithms to evolve. Whenever "ground truth" shows that the currently used algorithm does not provide satisfactory results, it may easily be changed or even replaced with an improved version. Thus the framework makes sure that it can always incorporate the most up-to-date state of science and technology.

Figure 10 shows the results of this process for the tree-wise delineation: Every tree receives its "digital business card", a collection of its properties that are relevant to its health, market value, treatment options etc.

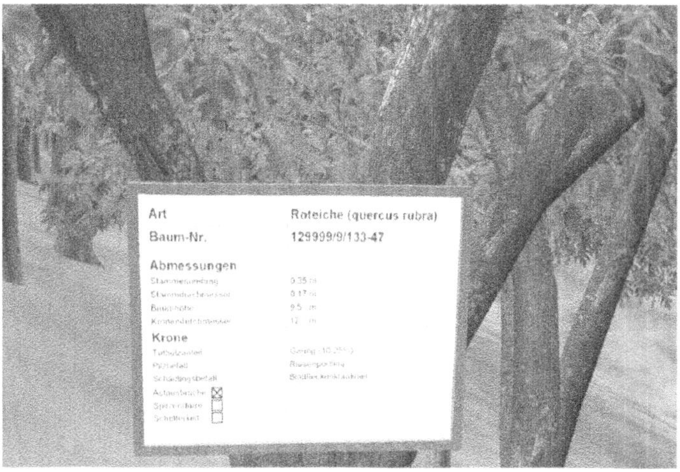

Fig. 10 The results of the remote sensing step are "digital business cards" for each single tree

One of the design issues — and the reason why different automation experts are involved — is the practical applicability of the generated tools, methods and data in a foresters daily work. As doing business in the forest is complex and has a manifold of complex decisions to make and complex tasks to perform, the Virtual Forest project is divided into 21 sub-projects, of which the most important will shortly be summarized in this chapter.

3.1 Forest Planning

In order to support forest planning, a new 4D-GIS approach was realized, based on an innovative virtual reality and simulation software. In general, the 4D (i.e. the 4 dimensional) aspect was of utmost importance, because the fourth dimension — the time — is an important property of all information saved in the Virtual forest database. Thus, all objects are not only known with their geo-coordinates, but all related information is also time-stamped, so that a user immediately gets an impression, how current the presented information is. Furthermore, the Virtual Forest "never forgets" which means that no information entered will ever be deleted: It may be superseded by more current information, but it will always be kept in a revision history that can be questioned to provide a "look back into the past" .

Being able to look back into the past is important in order to learn from the past, but combining this with capabilities to "look into the future" is the basis for economical success. Thus, the Virtual Forest integrates the tool SILVA [3], a single tree based stand simulator, to provide a "look into the future" concerning the development of the stand.

Figure 11 shows a sequence of pictures which resulted from the stand simulation with SILVA predicting the development of a stand within the next 5 years — including the thinning recommended to support elite trees. The sequence of images in Fig. 11 is way more convincing if viewed directly in the virtual world of the Virtual Forest, because one can "see the trees grow" in time-lapse mode and feels like being inside a "time-machine" .

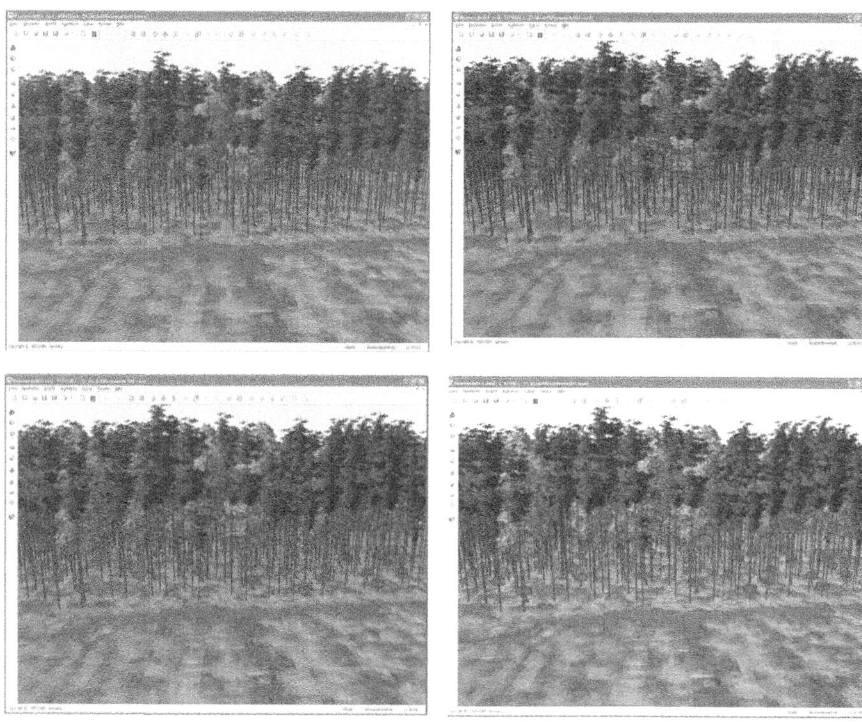

Fig. 11 From top left to bottom right — Results of the tree stand simulation with SILVA: The initial state of the stand, its state after 5 years of undisturbed growth, elite trees are chosen, support of elite trees through thinning

3.2 "Technical Production" in the Forest

Another important aspect of the virtual forest is the support of the "technical production" , i.e. the use of advanced mobile robotics technology in order to make the work of forest machines more efficient - e.g. by the introduction of a "automatic navigation to the next tree to fell" — feature — and also more environment friendly - e.g. by the support of automatic tire pressure control systems.

In order to be able to fully exploit the new capabilities of the Virtual Forest to support the technical production, i.e. the mobilization of wood resources, it is important that especially a harvester knows its exact location. If this is the case, the Virtual Forest's single-tree data, gained by the described remote sensing step, can be updated by incorporating the geo-referenced harvested trees: The harvester just has to "find out and remember" which tree was just cut and provide this information back to the Virtual Forest. In the Virtual Forest database, those tree are then treated as "cut" and so the calculation of the available wood volume can incorporate the latest state without waiting for the next remote sensing flight.

Furthermore, this approach resolves a problem typical of the state of North Rhine-Westphalia (NRW) with its huge number of forest owners with only small parcels. About two thirds of the privately owned forest in NRW consists of parcels of a size between 0.5 and 1.5 ha per owner. Having this new single-tree based bookkeeping available, the felled trees can exactly be assigned to the parcel where they were felled and thus to the owner.

Various experiments showed that the GPS accuracy under the tree canopy is not sufficient to be able to identify and distinguish between single trees. Thus, the Virtual Forest supports the equipment of harvesters with "optical GPS", a new technology that uses laser range finders (see Fig. 12) in order to get a measurement of the trees close to its position. It then uses the single tree positions, previously generated for the Virtual Forest through remote sensing as a map to localize itself. This is pretty much similar to human behavior: Look for landmarks, compare their configuration to a map, determine your position in the map.

Fig. 12 Machine logistics, deployment and chaining in the Virtual Forest: The forest machines are equipped with new sensors to enhance effectiveness

First tests of this approach really provided amazingly good results. In a tree stand of spruce, age 80 to 100, the harvester was able to localize itself with an accuracy of approx. 30 cm, which is way sufficient for the bookkeeping as planned.

The next step is close: A navigation system for harvesters which guide the harvester to the next tree to fell. It is already possible today, to use computer

Fig. 13 "Virtual fences" are metaphors for parcel boundaries, working paths can be projected onto the terrain in order to evaluate potentially difficult working conditions

software to automatically determine the trees to cut, e.g. in a thinning operation. With the harvester navigation system, the chosen trees don't have to be marked with paint any more, their positions will just have to be downloaded into the navigation system (see Fig. 14).

Beyond the advantages described above, equipping a harvester with the laser sensors also provides important feedback for the remote sensing procedure: The remote sensing procedure derives, e.g. the breast height diameter from the "visible, measurable or known" parameters of the tree like its height, crown diameter, age, water and nutrient supply etc. This estimated breast height diameter can now be compared to the breast height diameters the laser sensor "sees" and thus serve as a basis to further improve the derivation model. Thus, "ground truth" for the Virtual Forest is provided for free, as soon as the first thinning takes place.

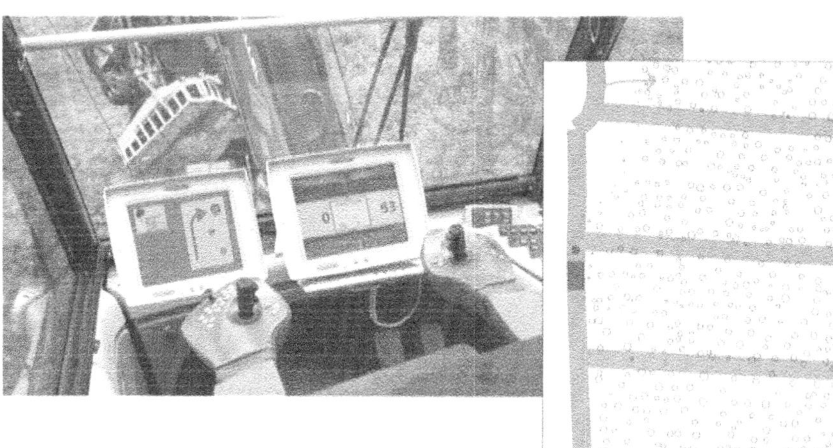

Fig. 14 "Harvester navigation to the next tree": Is this the coming generation of harvester navigation systems which directly guide the driver to the next tree to cut. Look at the left onboard display!

3.3 Advanced Working Machine Simulation

The Virtual Forest also drives forest machine simulation technologies to new heights - and to new application fields. A group of Virtual Forest partners around Prof. Warkotsch are currently developing a forest machine simulator that can not only be used for machine driver training, but also serves as the basis for the economical simulation of forest machine operation (Pausch 2005). The system allows to calculate the expected net profit by determining the logging cost with the help of an advanced simulation system. The determination of the logging cost can be performed very accurately today, because the approach carries out the logging in the virtual world as a simulation first and thus can calculate very precisely the logging cost to expect with respect to fuel, driving speeds, driver skills, machine capabilities etc.

Fig. 15 Advanced forest machine simulation technology is becoming more powerful but cheaper because the systems are now multi-use capable

As a side effect, this has opened a new application niche for the underlying simulator which originally was developed and marketed mainly for driver training. This new application opened new fields of application and lowered the introductory price for the driver training simulator to around 10.000 € which is almost an order of magnitude below current market prices and should further help to introduce latest virtual reality technology into the forest.

Last but not least, these developments open the door for a new quality of environmental simulations. As now remote sensing data can be gathered on a rather large scale — and at sensible costs — the new field of "semantic world modeling" opens up. The aim is here to get a better understanding of the complex processes in nature — and better monitoring tools to understand man-made disturbances in the fragile equilibra. The Virtual Forest can thus also be understood as a "Virtual Testbed" for new approaches to environment protection strategies.

Fig. 16 Close-to-reality simulation of the forest as a habitat for all kinds of trees, plants and wildlife

4 Conclusion

This paper explained how latest virtual reality, control and simulation technologies originally developed for space applications were combined to build high performance simulation and training "virtual testbeds" - also for the education of astronauts and industry workers as well as construction machine and forest machine drivers. For the realistic behavior of the simulated devices and machines, robotics knowledge is employed to model and to simulate the relevant processes physically correct and also to provide a correct kinematic and dynamic simulation of the vehicles involved. These capabilities were integrated with the virtual reality system CIROS® so that the trainees from all application fields easily get the impression to immerse into their virtual world for the respective training activities. It was also shortly described that the underlying software concept of CIROS® supported this application further in the way that it today also runs on large panorama projection screens or cave visualization structures in order to further improve the feeling of realism. The described simulators today are available as commercial products to help trainees to learn how to work safely, efficiently and ecologically sensible.

In addition to the more technology oriented testbeds, the Virtual Forest is about to become one of the most comprehensive and most current collections of knowledge about the forest and natural resources available today. Thus, it has the potential to serve as the basis for economical success in the forest as well as for new scientific results related to environment protection.

Thus, experiences and virtual reality components developed originally for space robotics testbeds have just opened up the door to a completely new field of applications.

References

[1] Rossmann J, Sommer B (2008) The Virtual Testbed: Latest Virtual Reality

Technologies for Space Robotic Applications. In: CDROM-Proceedings of 9th International Symposium on Artificial Intelligence, Robotics and Automation in Space (i-SAIRAS 2008), February 25–29, Los Angeles, CA, pp 78 ff

[2] Rossmann J (1999) Projective Virtual Reality: Bridging the Gap between Virtual Reality and Robotics, IEEE Transaction. on Robotics and Automation, 15(3):411–422

[3] Pretzsch H, Biber P, Dursky J (2002) The single tree based stand simulator SILVA. Construction, application and evaluation, Forest Ecology and Management, 162: 3–21

[4] Wokke F J P, Pronk Z (2000) Mission Validation and Training Facility for the European Robotic Arm (ERA). In: Proceedings of SESP 2000, 6th International Workshop on Simulation for European Space Programmes, by ESTEC, Noordwijk, The Netherlands, 10–12 Oct. 2000

[5] Argüello L, Alyoshin A, Fehse W, Miró J, Chliave P, Vankov A (1999) ATV Proximity Operations Validation & Rehearsal Using Distributed Interactive Simulation. In: Proceedings of AIAA International Space Station Service Vehicles Conference, Houston, Texas, Apr. 1999

[6] Piedboeuf JC, Carufel J de, Aghili F, Dupuis E (1999) Task verification facility for the Canadian special purpose dextrous manipulator. In: Proceedings of the International Conference on Robotics and Automation, Detroit, pp1077–1083

[7] Papasin R, Betts B J, Mundo R Del, Guerrero M, Mah R W , McIntosh D M, Wilson E (2003) Intelligent Virtual Station. In: Proceedings of 7th International Symposium on Artificial Intelligence, Robotics and Automation in Space, Nara, Japan

[8] New Man-Machine-Interaction concepts: Virtual Reality meets Robotics. In: Proceedings of the EUROBOT Conference 2008, Springer, pp89–96

Author

Prof. Dr.-Ing. Jürgen Roßmann, 2006-present, Chair and Director, Full Professor, Institute of Man-Machine-Interaction, RWTH Aachen University, Aachen, Germany; 2005–2006, Managing Director of EFR-Systems GmbH, Dortmund, Germany; 1998-present, Visiting Professor, University of Southern California, Los Angeles, USA; 1992–2005, Postdoctoral researcher and group head, Institute of Robotics Research, University of Dortmund, Germany; 1994, Visiting Scholar, University of Southern California, Los Angeles, CA, USA; 1991, Visiting Scholar, University of Texas at Austin, Austin, TX, USA. 2002, Habilitation for Control Systems in Automation, University of Dortmund, Germany; Habilitation Dissertation on Projective Virtual Reality; 1992, Doctor of engineering degree Department of Electrical Engineering of the University of Dortmund, Dortmund, Germany; Doctoral thesis "Collision Avoidance for Multi-Robot-Systems". Research interests are projective virtual reality, multi-agent supervision and control, action planning, multi-sensor-integration, digital factory applications, virtual human, system simulation techniques, real-time visualization,opti mization techniques, computer vision, man-machine interaction, teleoperation, medical applications.

Corresponding author:

Jürgen Roßmann(✉)
Institute of Man-Machine-Interaction, RWTH Aachen University, Ahornstr. 55, 52074 Aachen, Germany
E-mail: rossmann@mmi.rwth-aachen.de

Collaborative Virtual Assembly Operation Simulation and Its Application

Dengzhe Ma[1,2], Xijin Zhen[1], Yong Hu[1], Dianliang Wu[1,2], Xiumin Fan[1,2] and Hongmin Zhu[1]

1. CIM Research Institute, School of Mechanical Engineering, Shanghai Jiao Tong University
2. Shanghai Key Lab of Advanced Manufacturing Environment

Abstract

Nowadays complex products such as automobiles and ships are usually developed at geographically dispersed locations, which bring the difficulties for component model verification and assembly process evaluation. Collaborative virtual assembly (CVA) environment with VR technology is a good solution. However, the existing collaborative environments have many disadvantages in data management, reliability, function, etc, and furthermore they are difficult to meet the higher requirement of the users. In this paper, with the aim of solving the problems of model verification and assembly evaluation, two implement schemes for CVA system are analyzed, key technologies are discussed and prototype systems are developed. Finally, some applications of CVA are given to verify the scheme above.

Keywords

Collaborative Simulation, Parallel Rendering, Product Development, Virtual Assembly, Virtual Realty

1 Introduction

Complex products, hereto referring to the products such as automobiles, ships are complicated not only for function but for model structure; therefore, it is quite difficult for a single manufacturer to complete the entire development task for this kind of products; whereas the network-based colflaborative product design reflects the trend for product development. Usually, complex product development in a large-scale comprehensive enterprise involves lots of departments or suppliers. It is said that nearly 50% to 80% components of a complex product are supplied from collaborative partner manufacturers geographically dispersed [1], which means complex product development process needs more and more cooperation among designers. In this case, the lack of effective real time communication will bring difficulty for component model verification and overall product assemble-ability evaluation, and affect the whole product quality. On the other hand, product models become extremely complicated and refined for high-quality images requirement, as a result, a single PC cannot satisfy the demand of model rendering and other computing processing, especially a serious image delay occurring for the whole product. At present, there are mainly two methods to solve the problems above, namely, model simplification [2], exerting its influence by reducing the product data and the application of supercomputer, such as HP and SGI workstation [3], which can meet the requirement, but can't be popularized for expensiveness.

Collaborative virtual assembly (CVA) technology provides real time experimental assembly environment for designers located at different places, in which designers can share their product data, discuss and verify the assembly scheme to realize the planned design scenario. A real-time synchronized collaborative design process is a convenient way to express the design idea, and it brings great significance for improving design quality, shortening assembly evaluation time, and remote design guidance; however, the collaborative virtual assembly could not yet meet the operation simulation requirement of the complicated products in many aspects.

Grid technology has been applied successfully in areas of computer science requiring massive computing such as parallel computing, massive data process, and less in the area of design and manufacturing especially need real-time computing. However, the characteristics of GRID, service and resource sharing, collaboration, resource dynamic configuration, are just the requirements of CVA system. And then collaborative virtual assembly based on grid technology have many advantages in computing capability, data security, stability, scalability, etc, and furthermore they are easy to make applications for enterprise.

2 Related Works

2.1 CVA

With the increasing demand for product collaborative design and the development of network technology, collaborative virtual assembly has become an important research area. Lots of researchers have already conducted extensive research, and significant results have been achieved. Shyamsundar and Gadh [4, 5] studied the internet-based collaborative assembly and developed an internet-based collaborative product assembly design (cPAD) tool. The architecture of cPAD adopts 3–tier client/server mode. In this system, a new Assembly Representation (AREP) scheme was introduced to improve the assembly modelling efficiency. Bidarra et al. [6] presented a collaborative framework that does support integrated design of parts and assemblies; in this framework, the team members can discuss the assembly design issues through a collaborative validity maintenance scheme including phone, chat channel, shared camera, etc. Chen et al. [7] introduced an Internet-enabled real-time collaborative assembly modelling system, named e-Assembly, which allowed a group of geographically dispersed designers to jointly build an assembly model in real time over the Internet. C. Lu et al. [8] developed a collaborative assembly design environment which enabled multiple geographically dispersed designers to design and assemble parts collaboratively and synchronously through the Internet.

The above researchers proposed the approaches to realize the collaboration among different designers, but the interactive modes supported by the CVA environment are not the effective or intuitive ways such as chat channel etc. Besides, the performance of the system especially supporting real-time assembly activity for complex product was not considered. In order to address the above problem, Marsh et al. [9] discussed the relative performance of different distribution strategies, which support collaborative virtual reality such as client/server mode, peer-to-peer mode and several hybrid modes. They proposed a hybrid architecture which successfully supported real-time collaboration for assembly. For supporting the interactive visualization of complex dynamic virtual environments for industrial assemblies, Wang and Li [10] presented a dynamic data model integrating a spatial data set of hierarchical model representation and a dynamic transformation mechanism for runtime adaptation to the changes and movements of the assembly structures; based on model above, the complexity reduction was accomplished through view frustum culling, non-conservative occlusion culling, and geometry simplification. But the collaboration mode supported by above model is not considered.

The research works above have shown that more and more attention has been focused on this area. As collaborative assembly design involves the synchronized operation among several simulation systems, and because of large parts number and complex assembly structure, the real time data transmission through network is the very uphill work and the management of multi-user's collaborative information is relatively intricate; besides, the interactive mode is not natural. Therefore, there are still a lot of technologies waiting for further research.

2.2 Grid

Grid technology, first proposed in mid 1990s, was primarily targeted to enable large scale, dynamic collaboration among participants from highly heterogeneous and distributed computing systems. Grid connects geographically distributed computers, including high performance computers, workstations and PCs, databases, and all other equipment in the network, so as to form a virtual high-performance computing environment that is transparent to users.

Grid technology has been widely used by various communities to develop large-scale grid-based projects for handling variety of applications. These projects are focused on different functionalities: some primarily involved in development of Grid-enabled technology, such as middleware and infrastructure, others concentrate on exploring and harness grid technology in the context of specific fields of scientific research. Currently, the Grid and its application are still in their developing stage and the research work on collaborative engineering design on Grid environment is quite limited. Access Grid [11] connects people and teams via the Grid to enable group-to-group interaction and collaboration. Advanced Virtual Prototyping Research Centre (AVPRC) [12], in UK is developing a collaborative interactive virtual environment for engineering design. This project uses Access Grid technology for facilitating collaboration and computational steering. GEODISE [13] reports the development of an engineering design search and optimization involving fluid dynamics, and is bringing the collective skills of engineers and computer scientists together. The G-Yacht project [14] involved the development of a service to give remote access to yacht CFD and analysis tools via a Grid portal, enabling globally spread design teams to interact efficiently and effectively. Xu et al. [15] reviewed the computer aided collaborative work from a grid-computing viewpoint, and pointed out the trend of network computing and outlined its implications and new requirements. Li et al. [16] presented the concept of collaborative design grid (CDG) for product design and simulation, and its corresponding architecture was set up based on Globus toolkit 3.0.

Currently, some Grid computing toolkits, such as Globus [17], Legion [18], and Simgrid [19], which provide basic capabilities and interfaces for communication, resource location, scheduling and security, primarily adopt this client-server architecture based on centralized coordination. These Grid computing applications employ client-server architectures, in which a central scheduler generally manages the distribution of processing to network resource and aggregate the processing results. These applications assume a tightly coupled network topology, ignoring the changing topology of the network and are suitable for large enterprises and long time collaboration. However, for middle and small enterprises (SME), they may change their collaborators frequently. Current Grid systems are inflexible and rigid in term of their interactions and interoperations, and are not suitable for this situation. Therefore, a Grid framework with flexible and agile communication mechanism to enable collaborative design and manufacturing is important.

3 Function Requirements of Product Collaborative Assembly Simulation

To cope with the problem of collaborative design for complex products, traditional collaborative design solutions mostly focus on the product data and information management during product R&D [20]. Unfortunately, such ways are time-consuming and error-prone [21]. In contrast, we target the distributed team design scenario involving all the designers and supervisors related. By enabling geographically distributed designers to assemble their individual designs together in real time, an immersing and intuitionistic environment should be provided. To make this possible, a set of system requirements should be incorporated in the development as highlighted below.

- Internet condition and platform neutrality. To allow geographically dispersed designers to collaborate over the internet, the system should support the Internet standards and protocols, such as TCP/IP. And since the computing platforms of different designers are not guaranteed to be identical, the system should support all heterogeneous platforms such as Windows, UNIX.
- Convenient data conversion. The product modelling environment for designers may be different, so data conversion for common CAD software such as UG, CATIA, PRO/E should be supported.
- Real time interactive assembly operation. The designer can manipulate the object such as grasping, moving, constraint confirmation and motion navigation, release and such actions should be reflected accurately in the virtual environment. Tool plays an important role in assembly process, so the manipulation of object can do by virtual hand or virtual assembly tools.
- Synchronization control. The consistence of the scene of all the system nodes should be maintained, which means when the scene of a node changes because of manipulation, the information must transfer to all other nodes and update their scene synchronously.
- Large data set real-time process. A single PC's rendering capacity is not often sufficient for complex scene, especially when rendering high-quality, three-dimensional computer graphic images. Thus, the system should support scene rendering with a PC cluster.
- High-efficient collision detection method. Collision detection is the basic function of the interactive virtual reality simulation, while a more efficient collision detection method is necessary for complex products.
- Multi-modal Interaction. Users interact with VE through multiple modalities for different hardware, such as the common way keyboard and mouse like in CAD, 5DT Cyberglove and FOB (flock of bird), Cyber Glove/Touch glove with haptical feeling and FOB.

According to the requirements above, the basic idea of system is shown in Fig. 1.

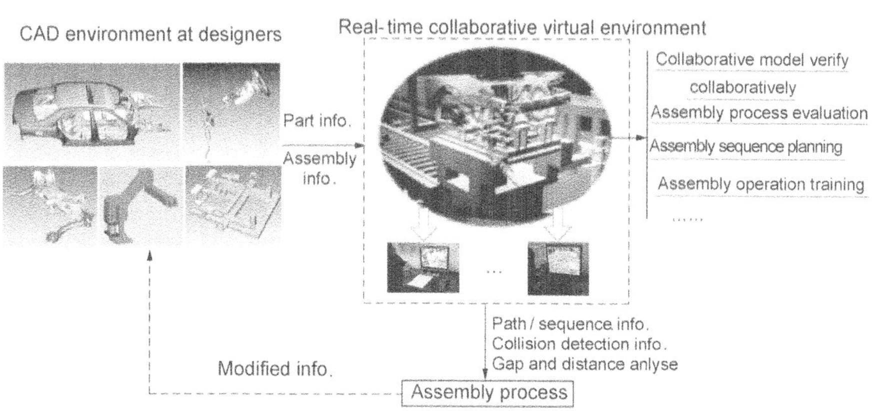

Fig. 1 Basic Idea of CVA System

Geometric modelling and assembly design of products are carried out in CAD system by designer at different locations. Then geometric and assembly information are transferred to system by a special data interface. All designers can work together to realize assembly analyses and assembly process planning of product collaboratively, as well as assembly operation training in the same immersive environment. The system can run in the internet or LAN (Local Area Network) condition.

4 Enabling Technologies of Collaborative Virtual Assembly

To realize multi-user collaborative virtual assembly, several enabling technologies need to be applied including virtual reality supporting platform, collaborative virtual assembly modelling technology, product data pre-process and management, multi-user collaboratively interactive operation technology.

4.1 Virtual Reality Supporting Platform

4.1.1 Large Scale Scene Rendering

The scene of complex product containing more complex, high-quality images and higher frame rates is required, and the rendering process becomes computationally demanding. A cluster composed of a collection of independent and cheap machines, used together as a supercomputer, provides a well-applicable and economical solution [22, 23]. A parallel rendering scheme for complex scene based on PC cluster and high-speed LAN is proposed. The key technologies include model segmentation and load-balance based on dynamic-frustum-division.

(1) Model segmentation for parallel rendering

The goal of polygon model segmentation is to avoid the situation in which big primitive sets crossing over several frustums and will be assigned to multiple renders for rendering; whereas repeated rendering will lead to efficiency loss. The flow chart of polygon model segmentation method is shown in Fig. 2 (a). Firstly, original polygon models are parsed into user-defined model tree by octree-spatial-division algorithm. If the number of primitives inside a partitioned model node is larger than a given number, the sub-node will be divided iteratively according to octree-spatial-division algorithm, until it is less than the given number.

(2) Load-balance based on dynamic-frustum-division

For the circumstance of imbalance distribution of screen projection of polygon models during the rendering process, a dynamic division algorithm is developed to perform frustum division. The flow chart of load-balance algorithm is shown in Fig. 2(b). The total-frustum is partitioned by recursive dichotomy method which depends on weight of render's capability. At present, there is not a good method to measure and acquire the value of rendering capability; therefore, the weight value is set to 0.5, which means all of renders have equal rendering capability, and only the movement of graphical primitive is under consideration during the frustum dynamic partitioning.

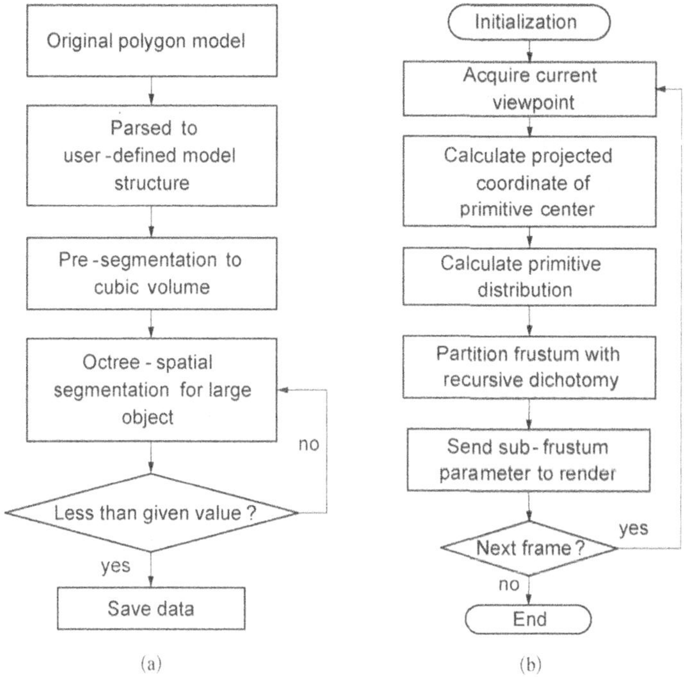

Fig. 2 Flow chart of Model Segmentation and Load Balance Process

(3) Two solutions for parallel rendering

a) Sort-first mode

Figure 3 is an improved sort-first system and the methods above are applied. The system is composed of a master node, several rendering nodes and several image processing nodes, which connect with high speed LAN. The number of supporting

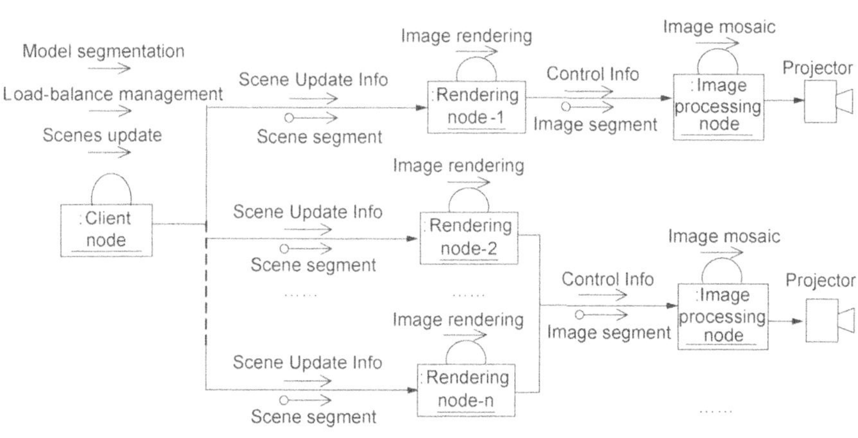

(a) Architecture of parallel rendering

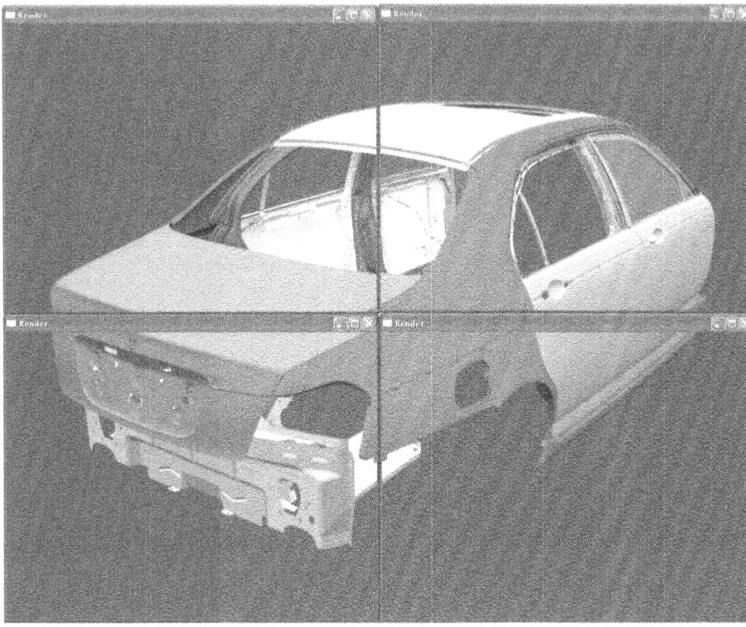

(b) 4-channel image from 4 rendering nodes

Fig. 3 Parallel rendering architecture and multi-channel implementation

nodes is configured by user at master node according to the performance of pc and the rendering task. Rendering nodes transmit result images to corresponding image processing nodes, where images are merged together to form a complete image, and after some necessary further processing such as geometry rectification and so on, the whole image will be output to display system, as shown in Fig. 3(a). With several image process nodes, multi-channel stereo (active and passive) can be realized. Figure 3(b) displays 4-channel car body image from 4 rendering nodes.

b) Sort-last mode

The framework is based on an improved sort-last structure, in which the large complex scene was split into N partitions, and one partition was dispatched to one render. The renders exchange its resulting image and depth with the other one after rendering, and then perform parallel pipeline image compositing with depth. Finally, the image processing node collects all the composited image partition to an integrated image. The image would be processed like rectification in succession. The result image would be output to display system in the end.

Parallel rendering system is composed of a scene management node, an image processing node and some rendering nodes, and these nodes are linked by high LAN of Gigabit bands as shown in Fig. 4. The number of nodes in simulation is determined by the performance of rendering node and the scale of rendering scene, which must be assured that users can be interactive with simulation system smoothly.

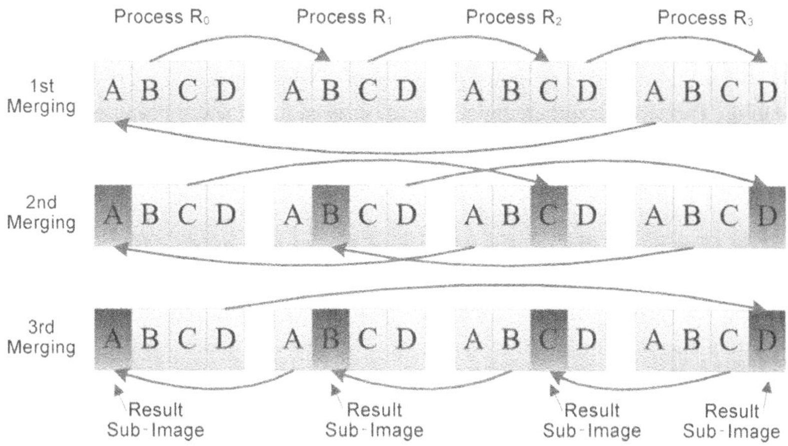

(a) Architecture of parallel rendering

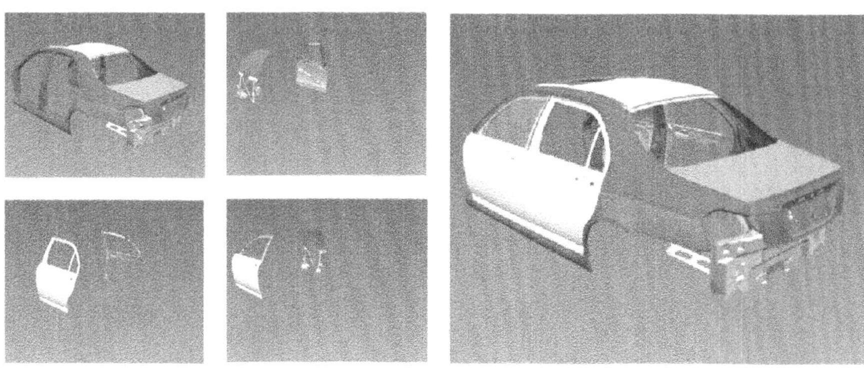

(b) 4 –channel image from 4 rendering nodes with sort-last mode

Fig. 4 Parallel rendering framework and virtual environment

4.1.2 High Effective Collision Detection Method

We present a new efficient CD algorithm for interactive virtual assembly. It is a pseudo-dynamic CD method based on bounding volume hierarchy (BVH) model of AABB trees, and adopts parallel computation method. It extends current techniques by simple but effective means in building the BVH models and CD. Our major contributions are:

New computation algorithms for generating BVH models of axis align bounding box (AABB) trees with different precision from general polygonal models.

A new algorithm for fast overlapping checking between two BVH models based on layered intersection checking method for AABBs.

(1) Fast BVH model generation from polygonal model

During pre-process, BVH models should quickly generate the simple polygonal models with the precision being controlled. Actually, polygonal models may have cracks, holes, T-joints or non-manifold geometry, which makes the pre-process time-consuming. Therefore, overlapping checking between a BV and a group of facets is the main time-consuming computing process. We present a simple method called "Point-In Test" (PIT) method, and combine it with "Separating Axes Test" (SAT) method to accelerate this process.

Basic principle of PIT is: if any vertex of a facet falls inside a BV, the facet must overlap with the BV; if not, whether they overlap with each other or not can not be determined. Compared with SAT, PIT method only needs some comparisons; therefore, its executing speed is much faster than DAC. As a result, whether a BV overlaps with a group of facets or not can be tested as follows: PIT is firstly applied to eliminate the facets overlapping with the BV definitely, and then SAT method is applied to test residuary facets.

Figure 5(a)–(d) show an example of BVH model generated according to the algorithm above.

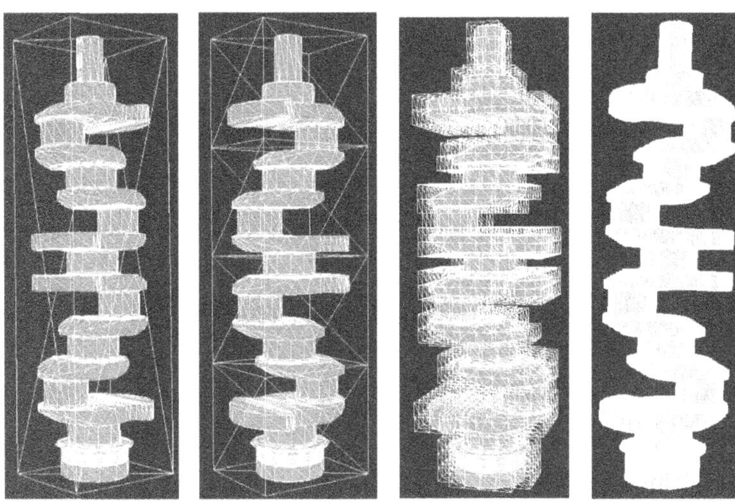

(a) Root BV (b) 200 mm (c) 5 mm (d) 1 mm

Fig. 5 BVH models with different precisions

(2) Fast CD between BVH models

In interactive operation process of VA, objects can be translated or rotated optionally. The AABBs in BVH models are changed into OBBs in virtual environment coordinate. So the overlapping test algorithm between two OBBs is the base of fast CD between BVH models. We represent "Nominal Radius Test" (NRT) and PIT methods as assistant algorithms for SAT to solve this problem.

For a bounding box, the radius of its maximum inscribed sphere is defined as its minimum nominal radius R_{min}, and the radius of its minimum circumscribing sphere is defined as its maximum nominal radius R_{min}. For two OBBs in virtual environment, supposing that d is the distance between their centers, and R_{max1}, R_{min1}, R_{max2}, R_{min2} represent their maximum and minimum nominal radiuses separately, the basic principles of NRT are:

If $d>R_{max1}+R_{max2}$, the two OBBs do not overlap with each other.

If $d<R_{min1}+R_{min2}$, the two OBBs overlap with each other definitely.

If $R_{min1}+R_{min2}\leq d\leq R_{max1}+R_{max2}$, the result can not be determined.

NRT method needs a calculation of distance and some comparisons, so it runs faster than SAT and PIT methods. Thus, to check whether two OBBs overlap with each other, the algorithms order applied is: firstly the NRT method should be adopted; secondly, when NRT fails, the PIT method should be used; finally, when both NRT and PIT method fail, the slowest SAT method should be used.

4.2 Collaborative Virtual Assembly Modeling Technology

4.2.1 Product Representation for Collaborative Virtual Assembly

In a distributed collaborative environment, the scene of all the users must be uniform. In addition, manipulation confliction resolution is another issue needed to take into consideration, requiring Product representation supporting collaboration and event synchronization mechanism.

Product assembly structure can be described in a form of assembly hierarchy tree, as shown in Fig. 6. Usually, product assembly is composed of sub-assemblies and parts, and sub-assembly is composed of parts. Constraint is used to describe the mating relation between two parts. Assembly, Part and Constraint are the primary object types in a collaborative virtual assembly system. Each object type has collaborative attributes which are used to express the object state, such as "Completed_Flag" of assembly representing whether an assembly is completed or not (when complete flag is set to "true", means assembly completed, vice versa.); "Position" of part referring to the coordinate location and orientation of an object; "Confirmed_Flag" of constraint meaning whether the constraint is confirmed or not (when the flag is equal to "true", means the constraint is confirmed, vice versa). If one user changes the scene, the modification of these attributes should feedback to other user in real time, and then the scene of each node must be uniform.

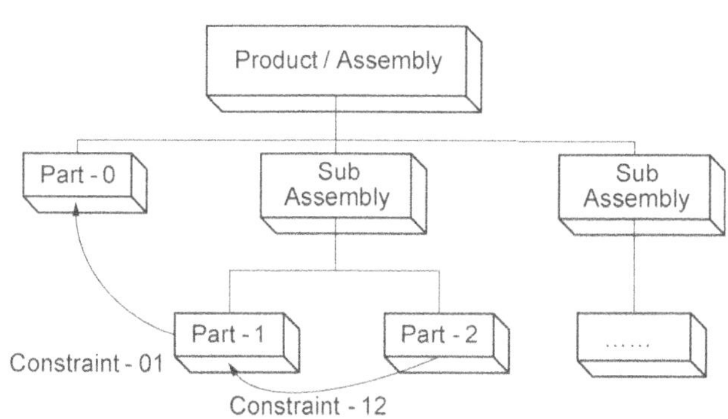

Fig. 6 Assembly hierarchy tree

Manipulation of an object (part or assembly) in collaborative virtual assembly process includes grasping, moving, constraint confirmation and motion navigation, release. In this process, the concept of attributes ownership is applied, which means node who owns the ownership of an object is responsible for updating the changeable attributes and broadcasting to all the others.

4.2.2 Virtual Assembly Tool Modeling and Operation

Considering the real assembly environment such as assembly tools, fixture, assembly equipment, is an important research content. Interactive assembly operation with assembly tool should be provided. In assembly process of complex product, equipments can be classified into three types: automation, semi-automatic, manual tool. There are many common characteristics in structure, using process and functions between tools and equipments (T&E), so we can use a unified method to model the information of tools and equipments. We treat tool and equipment as the same object, and apply OOP technology to represent a class for T&E with corresponding properties and functions.

In the viewpoint of model, just as showed in Fig. 7, a tool or equipment

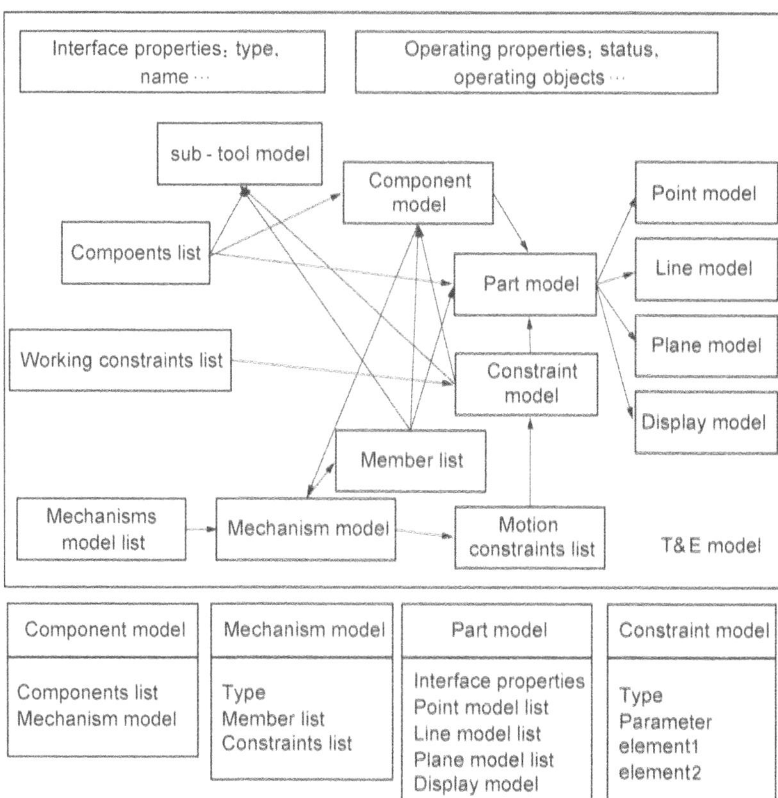

Fig. 7 Relationship among T&E model and its sub-models

model should contain lists of its sub-models, such as components list, working constraints list and mechanisms list. The element of components list can be a part model, a component model, or a sub-tool model. Where a component model can be composed of sub-components or parts, which can form internal mechanism models, a part model is composed of several point models, line models, plane models, and a display model. The working constraints list, holding the constraints between T&E and its operating objects, consists of constraint models, which can be defined between geometry elements from part models, component models, or sub-tool models. Mechanisms list, holding the motional mechanisms of T&E, consists of mechanism models, each of which contains lists of members and constraints between them. Where, the element of members list in mechanism model can be a part model, a component model, or a sub-tool model, the element of constraints list in mechanism model represents a motion limitation which is modeled as constraint model too.

Based on the analysis above, we can gain the classes of component mode, mechanism model, part model and constraint model, whose properties are listed at the bottom of Fig. 7. Where mechanism type in mechanism model represents the relative motion type between members, the interface information in part model includes its name, id, mass, material, etc; the information of constraint element1 and element2 should contain the corresponding interface information of parts they belong to.

4.3 Product Data Pre-process and Management

4.3.1 Acquisition and Pre-process of Product Data

Some prepared work should also be done for data transformation. Assembly hierarchy information and constraint information can be acquired in CAD environment after static interference check. Display model is "visible" part and collision detection model is used to check the dynamical interference. These models must be transformed from CAD models. A special interface based on ACIS solid modelling kernel has been developed for this complicated work. Each CAD model is transformed into .sat file, with which display model and collision detection model will be acquired:

- Display model: several types of models can be used such as .step, .flt (Open Flight), etc. All of them are polygon models with color, texture and other appearance properties. Several types can be supported by the system and all the models will be transformed automatically before the system starts.
- Collision detection model: this is a self-defined type transformed by a polygon model of last step, which is simplified into hierarchy bounding box according to collision detection precision [24].

All the data is saved into Grid data storage nodes by MYSQL system. When the system starts, all the computing nodes shall load the corresponding data.

4.3.2 Saving and Accessing Data

Data Management (DM) is one of the key factors, which provides a flexible and extensible mechanism for data manipulation. According to the system architecture and solution, users only need to upload CAD models related to the task to Grid data storage nodes. While in virtual environment, CAD model cannot be loaded directly; thus, some prepared work is required from CAD to virtual environment including data storage, process and access.

There are mainly three kinds of data for a virtual assembly task: product data including CAD model and data in virtual environment, information of assembly tools and process saving as a file, simulation result, as shown in Fig. 8. With the first type, product data in virtual environment includes display model, part information model, assembly hierarchy information (saved as file) and collision detection model. The second type is important auxiliary information for evaluation. Simulation result includes the data such as assembly process video, assembly sequence and path information file, evaluation report. Among these data, product data is managed and maintained by and only by task manager because of security who has the only authority to access database and others is only entitled to upload the data. The second type is shared by all the users of the task. Assembly result is saved at folder corresponding to users, which can be loaded by them with their authorities.

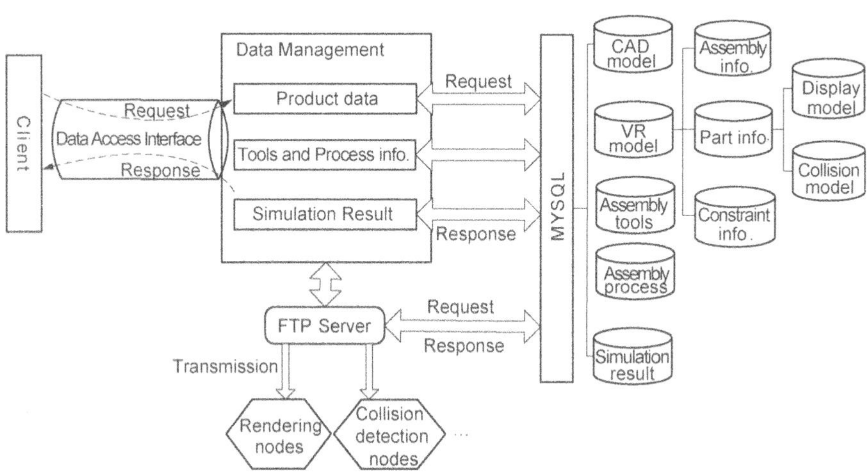

Fig. 8 Data management in VA-Grid

4.3.3 Data Protection

Sensitive information protection, such as appearance shape information and important geometric characteristics information, is an important problem in collaborative assembly process because the cooperation involves multi-users from different departments including component designer and supplier.

A product model can be transformed into several continuous multi-resolution models, users load different models into virtual environment according to their roles. Figure 9 demonstrates the original automobile defender model and two simplified models with few appearance shape features reserved. For example, a seat supplier can only own the second simplified models of defender. With the measures above, sensitive information can be protected effectively [25].

(a) Original model (b) Simplified model-1 (c) Simplified model-2

Fig. 9 Original automobile defender model and two simplified models

4.4 Multi-user Collaboratively Interactive Operation Technology

Unlike VA system, users at CVA system share the uniform virtual scene. They could feel the other users' operation at real-time and transfer operation to others. For this a virtual user model is present, which is a representation of real user. In this way, many users stay at assembly work station, observing with two eyes and operating with two hands; thus, all of them can feel each other without seeing.

4.4.1 Virtual User Models

According to system architecture, users could only send operation command and receive simulation result by network, with no product data saved or taking no computing task. The management of users and users' operation command is the very heavy work. So a virtual user model is used and each real user has a corresponding virtual user in virtual environment. The virtual user processes not only the current command of user but the attribution information such as location, viewpoint parameter. The partial model information of virtual user is shown as Fig. 10.

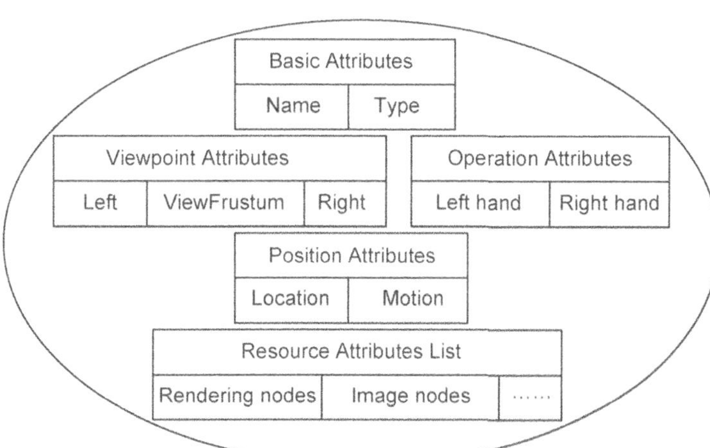

Fig. 10 Virtual user model

Here, "PlayerType" includes task manager and general collaborative user and "RNodeList" is the list of rendering nodes information service for this user. The system sets up a new virtual user object automatically when a client joins in.

In VA-Grid, there is no synchronizing problem among the clients because of one virtual scene, but manipulation conflict still exists:

- An object is grasped by two or more users simultaneously.
- Two objects that have assembly constraint relation are manipulated by two users.

For the former one, if the object has been grasped, it would refuse all, or continue. Then the authority and role of user are checked, otherwise, refuse him or accept his application. Finally, record the exact time and give the only right to the early user. For the latter one, firstly judge if it is a free part or not when a user tries to assembly one part to another; if yes, respond to assembly operation or continue to check. Then judge whether its operator is the same one as the user, if yes, it is the user who grasps two objects simultaneously, assembly can be continued with two-hand assembly process and if no, prompting message that the object has been operated will occur.

4.4.2 Remote Real-Time Interactive Assembling Operation

Many kinds of interactive operation equipment can be used such as FOB (Flock of Bird, a kind of position tracking device), data glove, mouse/keyboard and other I/O equipment. While in VA-Grid system, the data and its process program are all at the grid site, and the result at client is the rendering scene image. To realize interaction, the user's command must send to grid site in real-time. An image-based remote interactive scheme is provided, and the basic workflow is shown as in Fig. 11. The manipulating command is coded firstly and then sent to grid by network, which is decoded by system and sent to assembly simulation scene nodes. Then the parameter of related virtual user changes and all rendering nodes within "RNodeList" will be updated correspondently. In the same way, other computing nodes will update too, and new video figure will be gotten at client. Besides, users can communicate with each other by sending text message.

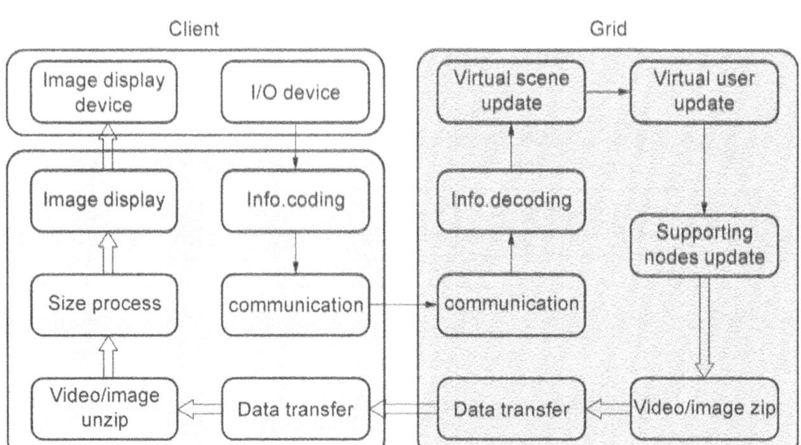

Fig. 11 Workflow remote real-time interactive assembling operation

By this scheme, user can manipulate the virtual object conveniently based on their hardware condition. The input and output is separated, making it getting multi-channel immersed stereo easily.

5 Applications of CVA System

5.1 Two Solutions of Collaborative Virtual Assembly

5.1.1 An HLA/RTI -based Collaborative Virtual Assembly

According to the system requirements and basic idea, an implementation scheme for collaborative virtual assembly based on HLA/RTI protocol[26] and MPI (Message Passing Interface)[27, 28] has been proposed as shown in Fig. 12. For this implementation scheme, there is a master node and several client nodes, the master and client are peer-to-peer node, who builds up collaborative virtual assembly federation together (concept in HLA). The master node is in charge of product data management and multi-user collaborative operation management. Each client node has the functions of virtual assembly simulation. According to the complexity of product model, the client node software can run in a single-PC, or in a PC-Cluster supported by high speed LAN. The amount of computers in a PC-Cluster can be configured based on the amount of polygon of product model.

With technology above, a prototype named DPVAE [29] is developed based on IVAE [30]. DPVAE is composed of six modules, namely, CAD system interface

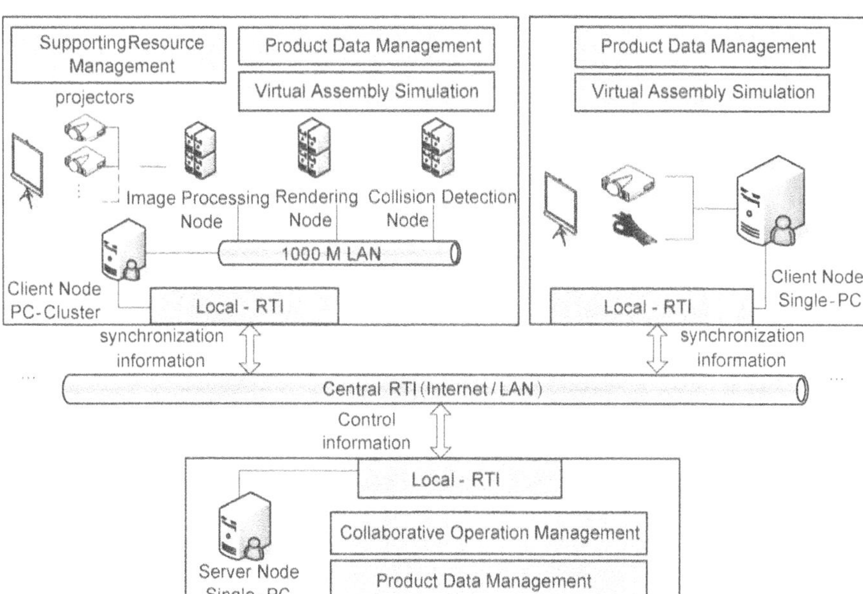

Fig. 12 Implementation scheme for collaborative virtual assembly simulation

module, master service module, client user service module, collision detection module, rendering service module and image process module, as shown in Fig. 13. In the system, product model from CAD environment can be imported to virtual environment by CAD data interface and models used for collision detection can be created based on part geometric polygon model. During assembly simulation, geometric models can be parallel rendered by rendering nodes and then output to image output nodes. After finishing the assembly operation task, an evaluation report that records assembly process information is created automatically. With these modules, DPVAE can be used to achieve assembly design and component model verification of automobiles collaboratively.

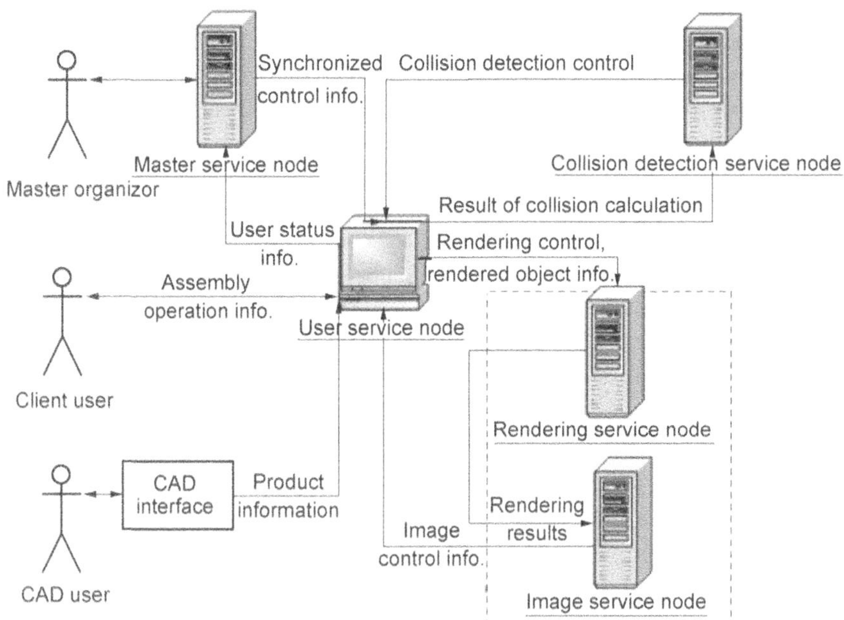

Fig. 13 System module structure of DPVAE

5.1.2 A Solution Based on Grid

An implementation scheme based on grid technology has been proposed as shown in Fig. 14 (Res.=Resource). There is only one virtual assembly scene in the system, maintained and updated by the grid. The computing resources needed during the assembly process such as rendering, collision detection, image process (may be the idle resource among the internet or intranet) are managed dynamically by the grid. Multi-task can be supported, and the product data and evaluation result related each task is stored at a relatively independent location. The configuration of user is simply, only I/O device or multi-channel stereo system if using immerse virtual environment. The scene at user-clients is a scene segment that a user can "see" at his location. The users are classified with assembly task manager and participants.

Another prototype VA Grid is also developed, which contains three main parts: portal, client and grid. Portal is the entrance of all the participants and client is the interface of interaction with the function of sending operating command and receives simulation result, as shown in Fig. 15.

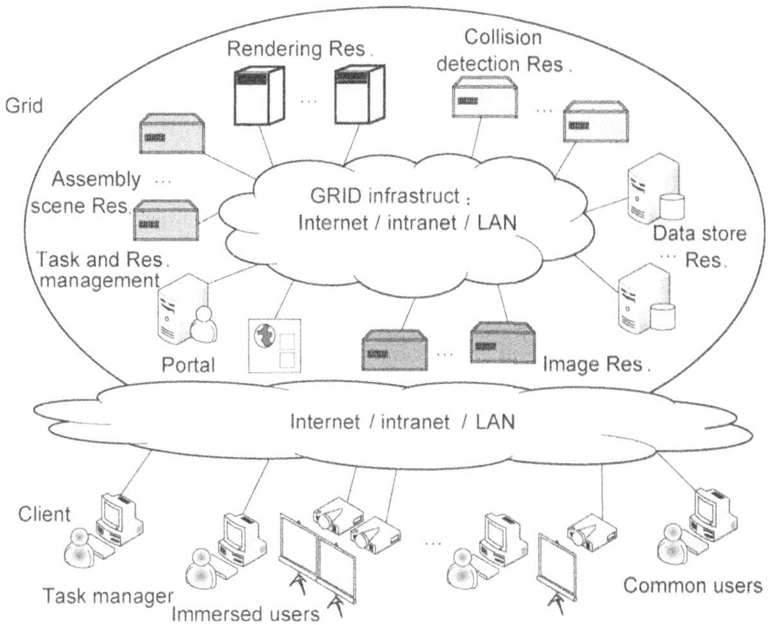

Fig. 14 CVA scheme based on grid

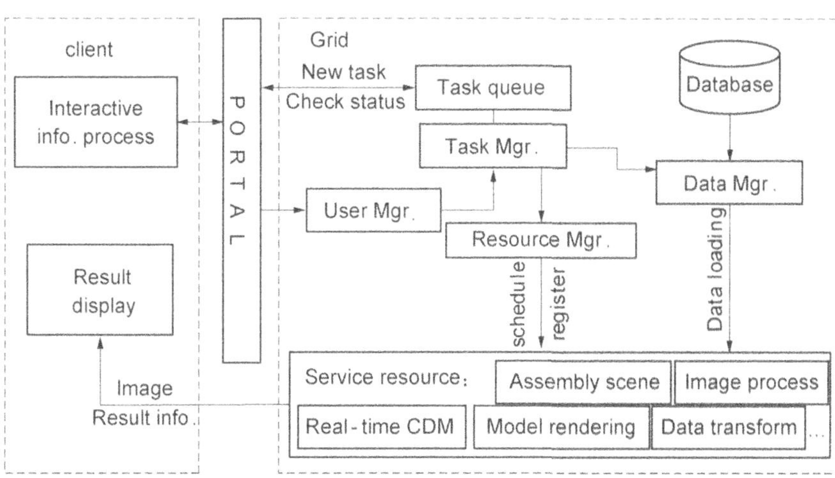

Fig. 15 System module structure of VA Grid

5.2 DPVAE of Application in Car Model Collaborative Verification

XX CAR is a new car and composed of 86 components, 396 parts (main component excluding connecting piece, such as screw). All the parts have been modelled, which has 6.93 million facets in total. The static interference checking of final assembly state of each component has been carried out in CATIA environment. It is difficult for designers to perform assembly evaluation task of whole car because of taking long time and low frame rate (about 2~6F/S (Frame/ Second)). The assembly process evaluation of whole car would be performed in DPVAE.

In this case, the evaluation involves designers forming departments of car body, chassis, inner ornament and a seat supplier, working at client nodes. Each client node configures: 4-PCs as rendering nodes, 2-PCs as image nodes and 1-PC as collision nodes, and each PC equips with Pentium IV 1.86 GHz, 2.0 GB of RAM, the connection between client node and supporting node is 1000 M LAN. The detailed process using DPVAE for this assembly evaluation is shown as follows:

5.2.1 Product Data Transform from CATIA to DPVAE

The process of the data transform can be divided into three steps: ① Get car assembly information from CATIA by the interface, which is developed with the RADE (Rapid Application Development Environment) and CAA/API (Component Application Architecture/Application Program Interface). ② Transform the CAD model to polygon model and collision model, whereas the former is used to display in virtual environment and the latter is used to be perceived and manipulated, the polygon model can be simplified as mentioned in part 5.2. ③ Put the polygon models and collision models into a given folder, and then write the saved folder path together with the car assembly information into a simulation file (.xml as given format), and all the user can load the models and the file easily by internet.

5.2.2 Create Collaborative Virtual Environment

A product manager is the supervisor and coordinator of the whole co-operating project, working at the master server node. The role and authority are defined according to their department. Then, the manager starts up the master server node with a given project name CASE1. Figure 16 shows the interface of virtual environment at server node. The master server node can monitor all users' status in real-time with user status dialog and exchange information with users by sending text message.

Designers firstly register the system and apply a role. The client node should be configured supporting resource nodes manually at first. Client node "VR-ZHEN" is in charge of car body design with two-channel stereo mode. Open the interface and select "Resource Monitoring", the resource nodes will be listed at the dialogue, as shown in Fig. 17. Browse the status of these nodes, select the idle node (the Utilization Ratio of CPU is lower than 40%) and add them as rendering service node (VR-R1, VR-R2, VR-R3, VR-R4), image service node (XZX, VR-HY) and collision detection service node (VR-GF). All the configuration information is saved in a file and then the system can be startup with "RUN", and the result of "VR-ZHEN" is shown as Fig. 18.

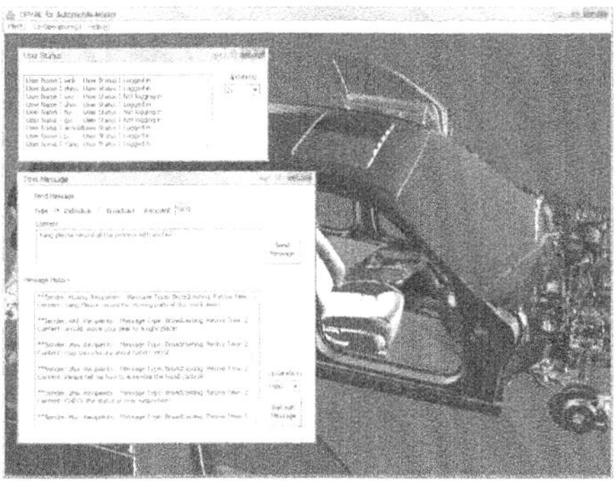

Fig. 16 Virtual environment at master node

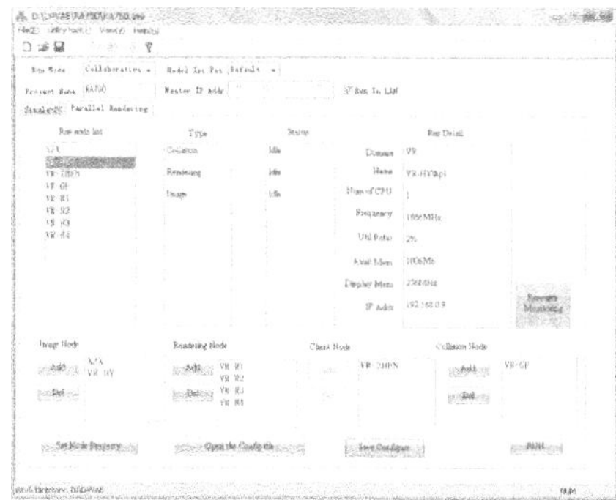

Fig. 17 Resource monitoring for parallel rendering client node

Client node "VR-YANG" is a designer for steering wheel with one-channel mode. The result is shown in Fig. 19. Client node "VR-ZHM" is a supplier for seat with one-channel mode, and the model at this node is the simplified ones except seat and related parts.

Fig. 18 Designer for car body (2-channel stereo)

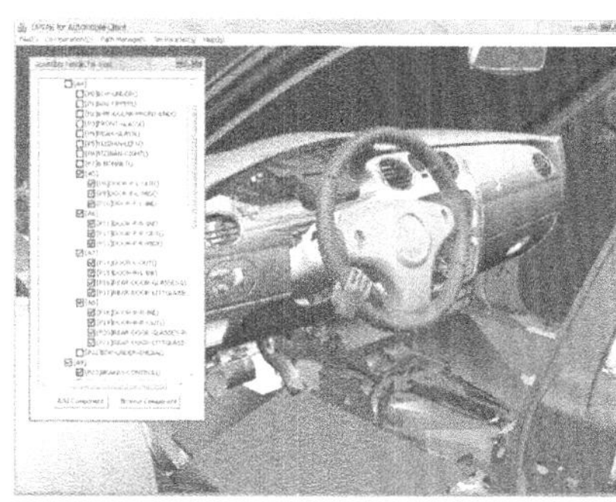

Fig. 19 Virtual environment at designer for steering wheel

5.2.3 Co-operating Assembly the Car in Virtual Environment

Under the Cooperative Control by the master node, users manipulate the component based on their authorities. From the components of chassis, car body to the doors and seats, the sequence should be kept as in real world, or repeated work (disassembly, reassembly) may occur. When a component is finished, its state will reflect at assembly tree dialog immediately with the checked box (as shown in Fig. 19). A user's operation can be displayed intuitively in other user nodes and server node synchronously. According to the problem during the assembly process, all the participants can communicate timely with each other by text message.

Each participant selects suitable interactive modes. Client node "VR-ZHEN" operate with Cyber/Touch data gloves and FOB (Flock of Bird) as shown in Fig. 18, "VR-YANG" with 5DT data gloves and FOB, "VR-ZHM" with the common I/O devices keyboard and mouse as shown in Fig. 20. The task can be achieved quickly and effectively because of the collaboration of all the users.

Fig. 20 Supplier of seats operating with keyboard and mouse

5.3 VAGrid of Application in Assembly Workstation Simulation

Aiming at verifying the feasibility considering real assembly space, assembly simulation is done with VAGrid system at classical workstation of rear suspension and former suspension. The content includes layout of fixture/tools, operational space, and the dynamic interference during assembly process, etc.

To use resource among internet, plug-ins must be set up at each resource node. In this case, computing resource among enterprise intranet is used, and product data is saved at inner database of enterprise. The detailed steps using VAGrid is shown as follows:

5.3.1 Register and Logon for Users

Each user firstly registers and gets an account. The task manager setups a simulation task team, defines roles and authority related this task. Figure 21 shows an interface that a user registers by portal.

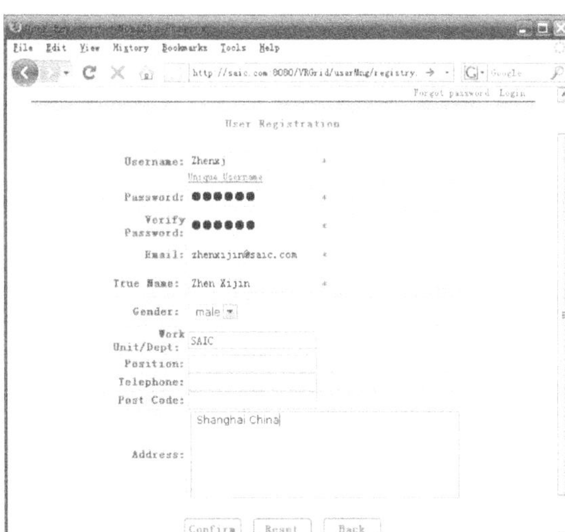

Fig. 21 User registers by portal

5.3.2 Preparation for Simulation Initialization

Three steps are used for Simulation Initialization: (1) All cad models are uploaded to grid database by designers from dispersed location; (2) Task manager access database and transform the models for simulation, and then saved to given location; (3) Write the saved folder path together with the car assembly information into a simulation file (.xml as given format). The segment of interface of transformation is shown as Fig. 22.

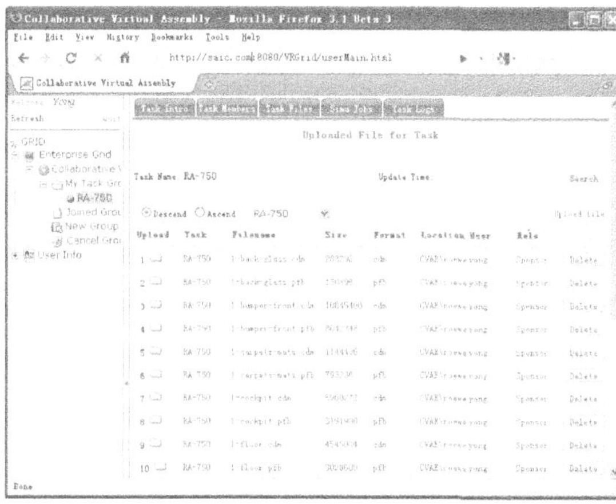

Fig. 22 Data process interface

5.3.3 Simulation Task Initialization

Task manager starts up the task with the file above and all related service nodes run automatically to support this task: assembly scene resource, rendering resource, etc. When all the related resource run, the status of this task is set as a collaborative task and users can join in. The whole assembly scene (at a default viewpoint) can be browsed by opening interactive interface, as shown in Fig. 23 , which is a part of assembly scene at task manager. Other users can apply a role and join in the task.

Fig. 23 Part of assembly scene

5.3.4 Multi-user Collaborative Assembly Operation

When all the related users join in the task, collaborative assembly operation will be done under the coordination of task manager. Several interactive models can be supported by the system, such as the common way keyboard and mouse in CAD, 5DT Cyber Glove and FOB (flock of bird), Cyber Glove/Touch glove with haptical feeling and FOB, as shown in Fig. 24, (a) user operating with keyboard and mouse; (b) operating with Cyber Glove/Touch glove and FOB.

Tools and equipment are important verifying objects in this case. Users can assembly by virtual tools and equipments by input the commands.

(a) Assembly spring with keyboard and mouse (b) Front suspension workstation

(c) Assembly with semi-automatic tool (d) Assembly rear suspension with equipment

Fig. 24 Multi-user collaborative assembly operation

6 Conclusion and Future Work

The collaborative virtual assembly environment is a useful computer-aided tool for supporting complex product design with which each designer can bring into exert their special advantages, communicate with each other through a more intuitionistic way, realize and solve the problems collaboratively in design, process, layout, etc. In this paper, two implemented schemes are given to provide a real-time collaborative design environment, which can solve the problem of massive data process and real-time interactive assembly operation, which meet the needs of collaborative virtual assembly.

By comparison, the scheme based on HLA/RTI has some disadvantages such as: ① data protection is difficult to realize for Data distribution; ② configure the supporting resource nodes manually and all the resource nodes is static and not reliable; ③ only can support a task at a time, not suitable for popularization. Its advantages are: ① rapid running speed especially at initialization because data is saved at local position; ② only common network condition is need.

However, VAGrid system can fully use the idle resource among the internet and simply configuration requirement for users. Data protection can be done successfully. But this system relies on good network and plug-ins are needed for resource nodes.

Future work is to resolve the shortcomings of two implemented schemes and improve system operating efficiency, and carry out man-machine ergonomics applied research, etc.

Acknowledgments

This work is supported by the Key Project from National Nature & Science Foundation of China (Grant no. 90612017) and Key Project from Science & Technology Commission of Shanghai Municipality (Grant no. 08DZ1121000).

References

[1] Rezayat, M (2000) The Enterprise-web Portal for Life-cycle Support. Computer Aided Design 32(1):85–96

[2] Li W D, Cai Y L, Lu W F (2007) A 3D Simplification Algorithm for Distributed Visualization. Computers in Industry 58(3):211–226

[3] Iyer R, Perdue J, Rauchwerger L, Amato NM, Bhuyan L (2005) An Experimental Evaluation of the HP V-class and SGI Origin 2000 Multiprocessors Using Microbenchmarks and Scientific Applications. International Journal of Parallel Programming 33(4):307–350

[4] Shyamsundar N, Gadh R (2001) Internet-based Collaborative Product Design with Assembly Features and Virtual Design Spaces. Computer-Aided Design 33(9):637–651

[5] Shyamsundar N, Gadh R (2002) Collaborative Virtual Prototyping of Product Assemblies over the Internet. Computer-Aided Design 34(10):755–768

[6] Bidarra R, Kranendonk N, Noort A, Bronsvoort W F (2002) A Collaborative Frame-work for Integrated Part and Assembly Modeling. In: Proceedings of the 7th ACM Symposium on Solid Modeling and Applications, New York, pp 389–400

[7] Chen L, Song Z J, Feng L (2004) Internet-enabled Real-time Collaborative Assembly Modeling via an e-Assembly System: Status and Promise. Computer-Aided Design 36(9):835–847

[8] Lu C, Fuh J Y H, Wong Y S, Qiu Z M, Li W D, Lu Y Q (2006) Design Modification in a Collaborative Assembly Design Environment. Computing and Information Science in Engineering 6(2):200–208

[9] Marsh J, Glencross M, Pettifer S, Hubbold R (2006) A Network Architecture Supporting Consistent Rich Behavior in Collaborative Interactive Applications. IEEE Transactions on Visualization and Computer Graphics 12(3):405–416

[10] Wang Q H, Li J R (2006) Interactive Visualization of Complex Dynamic Virtual Environments for Industrial Assemblies. Computers in Industry 57:366–377

[11] Access Grid. http://www.accessgrid.org.

[12] Bee S, Ford R, Margetts L, Roy K (2003) Collaborative Virtual Design in Engineering. http://www.sve.man.ac.uk/General/Staff/margetts/publications/manchester_showcase.pdf. July, 2009

[13] GEODISE. http://www.geodise.org.

[14] G-Yacht. http://www.soton.ac.uk/gyacht/.

[15] Xu Z W, Liao H M, Li B C, Li W (2004) Vega Grid and CSCW: Two Approaches to Collaborative Computing. In: Shen W et al (eds) Proceedings of the 8th International Conference on Computer Supported Cooperative Work in Design, vol 2, Beijing, pp 10–17

[16] Li Z, Jin X L, Cao Y, Zhang X Y, Li Y Y (2007) Architecture of Collaborative Design Grid and Its Application based on LAN. Advances in Engineering Software 38(2):121–132

[17] Foster I, Kesselman C (1997) Globus: A Metacomputing Infrastructure Toolkit. International Journal of Supercomputer Applications and High Performance Computing 11(2)115–128

[18] Natrajan A, Humphrey M A, Grimshaw A S (2003) GRIDS: Harnessing Geographicallyseparated Resources in a Multi-organisational Context. In: Kent RD, Sands T W (eds) High Performance Computing Systems and Applications, pp 25–32

[19] Casanova H (2001) Simgrid: A Toolkit for the Simulation of Application Scheduling. IEEE Comput. Soc, Brisbane

[20] Trappey A, Hsiao D W (2008) Applying Collaborative Design and Modularized Assembly for Automotive ODM Supply Chain Integration. Computers in Industry 59(2–3):277–287

[21] Zhuang Y, Chen L, Venter R (2000) CyberEye: an Internet-enabled Environment to support Collaborative Design. Concurrent Engineering 8(3):213–229

[22] Liang K, Monger P, Couchman H (2004) Interactive Parallel Visualization of Large Particle Datasets. Parallel Computing 25(2–3):89–98

[23] Xiong H, Peng H Y, Qin A H et al (2007) Parallel Strategies of Occlusion Culling on Cluster of CPUs. Computer Animation and Virtual Worlds 18(3):165–177

[24] Wu D L, Zhu H M, Fan X M (2008) A Parallel Collision Detection Algorithm for Interactive Assembling Operation in Virtual Environment (in Chinese). Journal of Shanghai Jiao Tong University 42(10):1640–1645

[25] ZhenX J, Wu D L, Fan X M et al (2008) Research on the Technologies of Multi-users Modeling and Cooperation in Distributed Virtual Assembly System (in

Chinese). Journal of System Simulation 20:124–127

[26] IEEE Standard for Modeling and Simulation (M&S) High Level Architecture (HLA)—Framework and Rules (2000) Institute of Electrical and Electronics Engineers Inc.

[27] Le T T, Rejeb J (2006) A Detailed MPI Communication Model for Distributed Systems. Future Generation Computer Systems 22(3):269–278

[28] Message Passing Interface Forum (2008) http://www.mpi-forum.org

[29] Zhen X J,Wu D L,Fan X M et al (2008) Distributed Parallel Collaborative Virtual Assembly System for Complex Products (in Chinese). Computer Integrated Manufacturing Systems 14(10):1990–1995

[30] Yang R D, Fan X M, Wu D L et al (2007) Virtual Assembly Technologies based on Constraint and DOF Analysis. Robotics and Computer-Integrated Manufacturing 23(4):447–456

Author

Prof. Dr.-Ing. Dengzhe Ma is a professor of Shanghai Jiao Tong University (SJTU), the director of Shanghai Key Lab of Advance Manufacturing Environment and the previous head of Computer Integrated Manufacturing Research Institute of SJTU. He received Bachelor Degree in Mechanical Engineering from SJTU in 1964 and his Dr.-Ing. Title in Mechanical Engineering Department of Darmstadt Technology University in Germany in 1987.

Xijin Zhen is a Ph.D. candidate of Shanghai Jiao Tong University whose research interests are collaborative virtual assembly for product design and manufacturing.

Yong Hu is a Ph.D. candidate at Shanghai Jiao Tong University whose research interests are virtual reality platform technology, parallel rendering, and virtual reality grid.

Dr. Dianliang Wu is a lecturer of Shanghai Jiao Tong University whose research interests are virtual prototyping, virtual assembly, reverse engineering rapid prototyping and computer aided engineering. From 2003 to 2005, he was a post doctor fellow in SJTU. He received Ph.D. in Mechanical Engineering of Xi'an Jiao Tong University in 2003.

Prof.Dr.Xiumin Fan is a professor at Computer Integrated Manufacturing Research Institute, School of Mechanical Engineering, Shanghai Jiao Tong University. During the study period from 1989 to 1998 in Shanghai JiaoTong University, she received Bachelor (1993), Master (1996) and Doctor Degree (1998) in Mechanical Engineering Department. In 2004, she received Shanghai Women Innovation Nomination Award, and was selected as the Outstanding-young-teacher back-up candidate in Shanghai colleges and universities. In 2006, she received the 10th Ying Tung Fok Education Foundation Award for Young Teachers (research category). In 2007, she was selected to joint the Program for New Century Excellent Talents in University from The Ministry of Education. Her research interests are virtual prototyping and virtual manufacturing.

Hongmin Zhu is a Ph.D. candidate of Shanghai Jiao Tong University whose research interests are virtual assembly process planning, product model information enhancement for virtual assembly and collision detection algorithm.

Corresponding author

Xiumin Fan(✉)
CIM Research Institute, School of Mechanical Engineering, Shanghai Jiao Tong University, No.800, Dongchuan Road, Shanghai 200240, China.
E-mail: xmfan @sjtu.edu.cn

Integration of Realtime Ray Tracing into Interactive Virtual Reality Systems

Hilko Hoffmann[1], **Dmitri Rubinstein**[1], **Alexander Löffler**[2], **Michael Repplinger**[1] **and Philipp Slusallek**[1, 2]

1. German Research Center for Artificial Intelligence (DFKI)
2. Computer Graphics Group, Saarland University (UdS)

Stuhlsatzenhausweg 3, Campus D3.2, 66123 Saarbrücken, Germany

Abstract

Current processors provide high performance through parallelism by integrating more and more computational cores on a single chip instead of increasing the clock rate. This is true for both the CPU (multi-core of up to 8 cores) and even more so for the GPU (many-core of up to 240 cores). GPUs are still being programmed in vendor specific languages (like Nvidia's CUDA) but cross-vendor initiatives like OpenCL will allow for providing performance on a standard desktop PC that was previously only possible on supercomputers. With is upcoming Larrabee processor, Intel goes one step further and tries to combine the concepts and advantages of multi-core CPUs with that of many-core GPUs. It moves the entire rendering process into software providing more flexibility to realtime graphics applications like games or visualization applications.

In this paper we present a highly parallel System consisting of the completely new Realtime Ray Tracing engine "RTfact" and the Realtime Scene Graph "RTSG" that allow making good use of modern parallel hardware. RTfact accelerates rendering via ray tracing to the point where it can be used for interactive Virtual reality applications, while RTSG allows for flexible and high-level descriptions of 3D environments on the basis of the X3D standard that enable the description of 3D objects and their behavior. RTSG is thus the interface between Virtual Reality systems and a number of different rendering modules that includes ray tracing as well as fast rasterization via the OGRE library.

RTSG currently is the fastest X3D browser that optimally supports construction and design decisions through high image quality, exceptional visual realism, as well as the high degree of detail in scenes.

Keywords

Real-Time Ray Tracing, Virtual Reality System, Scenegraph, X3D, Many-core CPU

1 Introduction

One application area for Virtual Reality (VR) applications is the support for visual decision making for design, product development, as well as architectural visualization. Those use cases need the ability to accurately visualize the object properties such as the type of material, roughness, color, shape, or curvature. This in turn requires an accurate simulation of the illumination in an environment, possible from many light sources.

Current VR systems mostly use classical rasterization techniques to generate images. However, in contrast to computer games where the content development and optimization of visual display may require several man years, VR applications must be able to display and interact with a 3D scene almost immediately and essentially without much preprocessing or manual tuning of scene and display parameters. Even though, the results must abide to even higher standards, as the displayed images should not only look nice but are the basis for possibly far reaching decision — and therefore should be reliable and correct.

This is where traditional rasterization technology reaches its limitations, despite its high, hardware-accelerated rendering performance: The predefined graphics pipeline [1] makes it hard to flexibly use it also for other rendering techniques. Most of the available techniques put visual effects before physical realism, and the development effort as well as the rendering cost increases tremendously when realism for arbitrary scenes must be supported. A good example is Nvidia's demo "Medusa" [2], which requires a total of 120 separate rendering passes per frame.

Realtime Ray Tracing [3] is an alternative approach to rasterization that supports the physically correct simulation of illumination, reflection, hard and smooth shadows,

Fig. 1 Ray tracing of realistic reflections of glass and varnish surfaces

dynamic light sources, and measured materials and luminaires. The resulting images are of high visual quality and with good input data hard are hard to distinguish from reality (Fig. 1). However, ray tracing has very high computational demands that are not supported by specialized hardware.

2 The Realtime Ray Tracing Engine "RTfact"

Modern CPUs offer high performance through parallelism in the form of many cores and SIMD (Single Instruction Multiple Data) processing, which makes them well suited to accelerate ray tracing. New algorithmic approaches replace the predefined graphics pipeline by the flexibility and modularity of software that enables new kinds of visual realism. SIMD instructions are most effective when many similar computations have to be done simultaneously. We can exploit this by tracing an entire packet of rays together instead of individual rays because similar rays typically require very similar computations. However, the implementation of these advanced ray tracers is significantly more complex, these SIMD instructions must still be manually coded due to limitations in current C/C++-Compiler that have difficulties to automatically convert code to SIMD form. Instead, a programmer in the past had to code the core algorithms as well as the individual shaders in what is essentially assembly level programming using so called "instrinsics". Such ray tracing application is fast but inflexible, hard to extend, and largely non-portable to new processor versions. More flexible software approaches using common object-oriented design techniques and external libraries simplify programming but suffer significantly in performance.

The goal of the newly developed ray tracing engine "RTfact" [4] was to maximize performance on modern CPUs as well as GPUs without compromising flexibility. RTfact is no complete rendering system but a template C++ library that offers building blocks for the assembly of optimized and well adapted ray tracers. Through the use of C++ temples we combine the performance of SIMD code with the flexibility of object-oriented programming. However, in contrast we typically separate the algorithms from the concrete data representations that would normally put into the same class into separate template constructs ("concepts"), which follows the design principles used by the STL and Boost C++ libraries.

However, the result is no longer a binary ray tracing library but a set of source code files from which the compiler selects and combines the suitable features during compile time, based on very high level C++ template instantiations. This code is then automatically inlined into big basic blocks, which would be hard to write and maintain by hand but offer great optimization opportunities for a compiler. In order to still provide traditional binary interfaces and shared libraries (DLLs), we create preconfigured binary libraries that cover the most used and optimized applications scenarios. For other cases the developer can directly use the original templated interface.

3 The Ray Tracing Scene Graph "RTSG"

The ray tracing engine RTfact offers rendering functionality on a level comparable to OpenGL or DirectX. However, most applications including VR systems prefer to work on higher level of abstractions — so called scene graphs. They allow for organizing a 3D scene using a set of hierarchically organized elements (e.g. geometric objects, light sources, cameras, sensors, etc.) instead of working with low-level API calls. For OpenGL there exist a number of scene graph libraries, such as OpenSceneGraph [5], or OpenInventor [6].

The significantly different rendering approaches makes it largely impossible to integrate RTfact with any of them as they have been optimized for OpenGL-style rendering that is hard to emulate in a ray tracing engine. The resulting system would be too inefficient and slow to support realtime VR applications [7]. Instead we designed a new scene graph better suited also for ray tracing. The resulting library RTSG is based on the ISO standard "X3D" [8].

The current version supports both RTfact for ray tracing as well as the "Object-Oriented Graphics Rendering Engine (OGRE)" [9] which in turn uses OpenGL und DirectX for rendering. The programming interface of RTSG is based on the "X3D Scene Access Interface" (SAI) and is completely renderer agnostic. This level of abstraction also allows for creating hybrid rendering systems that simultaneously talk to multiple renderers and may even combine their results. No changes are required in the application as the rendering configuration can be specified in a separate application-independent configuration. Despite its separation between scene graph and rendering RTSG today is clearly one of the fastest X3D viewers available.

4 The Distribution Framework "URay"

RTSG and in particular RTfact are able to fully exploit the computational power of a PC but for very large scenes and highly realism the processors of a single PC may not be sufficient. In those cases it would be advantageous to be able to also exploit the capacity of other computers in the form of an on-the-fly or dedicated cluster of PCs. For that purpose we developed the "URay" framework [10] that can distribute and synchronize the computation across the Internet. This framework is based on the "Network-Integrated Multimedia Middleware (NMM)" [11] operating on distributed flow graphs (see Fig. 2).

The nodes of this graph represent individual processing steps (like ray tracing, tone mapping, or display) and can be distributed across multiple machines. For distributed rendering, a master node splits the frame buffer into tiles that are sent to a number of rendering nodes, which are typically separate machines so we can use all of their combined computational power. The resulting tile of pixels data are finally composited and displayed in additional nodes.

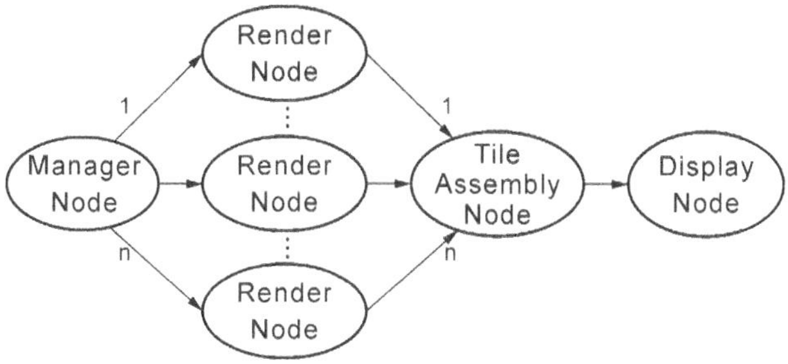

Fig. 2 Example of an NMM flow graph within URay

URay offers a large number of rendering and display options, including displaying the same image on multiple displays or tiling the display independently from the tiling used for rendering, or full stereo across all displays. Additional image processing nodes can be inserted in the flow graph to perform operations such as tone mapping, image warping, edge blending, and others. All displays can be synchronized across the network using NMM's built-in high-quality distributed synchronization framework. This general distributed rendering approach makes use of the fact that pixels are independent in ray tracing, which leads to a highly efficient and linearly scalable distribution approach.

5 Integrating Ray Tracing into a VR System

The software architecture of established VR systems, e.g., Lightning [12] or VR-Juggler, generally consists of three layers: an application and interaction layer (high-level), a scene graph system (mid-level), and a rendering layer (low-level). State-of-the-art systems based on rasterization mainly use OpenGL for low-level rendering libraries because of its platform independence and wide hardware support. The layer above, responsible for graphic abstraction and organization, usually uses classic scene graphs mostly on an OpenGL base, such as OpenSceneGraph or OpenSG. Only the application development on the top layer discriminates between the concepts of different VR systems, in this way defining their target group.

To integrate realtime ray tracing into an existing VR system, one can theoretically begin on any of the three layers. However, the application layer should remain unchanged because a change of this layer would affect all existing and future applications on the VR system and would necessitate writing them anew.

Using a rendering engine for realtime ray tracing inside the rendering layer instead of an engine based on OpenGL must fail in practice (as described in Section 3). The reason for this is that the structure of the scene graphs used is unsuitable for a reasonable integration of ray tracing [13]. Our approach within the VR system Lightning therefore uses the scene graph RTSG, which is completely independent of a specific rendering technology and supports both ray tracing and rasterization. RTSG already includes the ray tracing engine RTfact, as well as the rasterization engine OGRE.

The rendering specific components of the scene graph OpenSceneGraph originally used by Lightning are completely replaced by RTSG in our approach. Figure 3 shows the applied architecture and the different levels of integration.

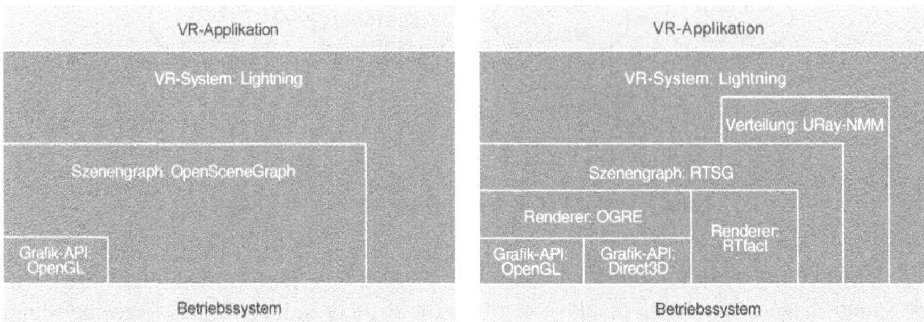

Fig. 3 Architecture of a classic VR system (left) and the integrated system (right)

RTSG implements the X3D standard, providing standardized interfaces for an easy integration into existing VR and visualization systems. X3D also provides interfaces for powerful languages for defining application logic, for instance by the *scene authoring interface* (SAI).

In a typical VR system, there are many different possibilities for the realization of application logic. For example, the VR system Lightning has a C++ and a TCL/TK script interface. The developer has to decide on which level applications or components of application should be defined.

The proposed approach separates the application logic into two fundamental areas: object related components that can be directly assigned to the behavior of an object within the scene (e.g., a traffic light pre-emption). Components spanning different objects define the behavior of the entire system (e.g., traffic simulation). The previous ones are implemented directly in X3D and can be reused in different applications, while the latter are developed on the level the VR system.

6 Conclusions and Future Work

Through the development of multi core processors that can be programmed flexibly, ray tracing is a genuine alternative to image rendering using graphic boards. In VR applications, ray tracing provides an inherent image quality and a degree of realism that the classic methods cannot achieve without enormous effort. Today's flexible hardware increasingly dissolves the borders between rasterization and ray tracing. This makes hybrid systems imaginable benefit from the advantages of both approaches.

The integration described above shows that the use of ray tracing in immersive, interactive applications with a justifiable amount of hardware. However, there is still a need for high performance clusters to achieve a smooth interaction and reaction of the system. The possibilities of visualizing applications with highly dynamic scene content are currently very limited, though the image quality is high.

The big breakthrough is probably only to be expected when ray tracing can be used by computer games on desktop computers and is integrated into common graphic drivers or development libraries, respectively.

References

[1] Akenine-Möller T, Haines E (2002) Real-Time Rendering. AK Peters, Natick, Mass. 2002

[2] Golem (2008) Nvidia Tech-Demo "Medusa". http://www.golem.de/0806/60425.html

[3] Wald I, Benthin C, Wagner M, Slusallek Ph (2001) Interactive Rendering with Coherent Ray Tracing. Computer Graphics Forum 20, 3: 153–164 Sept 2001

[4] Georgiev I, Slusallek P (2008) RTfact: Generic Concepts for Flexible and High Performance Ray Tracing. In: Proceedings of IEEE Interactive Raytracing Symposium

[5] Kuehne B, Martz P (2007) OpenSceneGraph Reference Manual v2.2. Skew Matrix Software and Blue Newt Software

[6] Wernecke J (1994) The Inventor Mentor. Addison-Wesley

[7] Dietrich A, Wald I, Wagner M, Slusallek P (2004) VRML Scene Graphs on an Interactive Raytracing Engine. In: Proceedings of IEEE VR, 2004

[8] Kloss J, Schickel P (2009) Programmierung interaktiver Multiuser-Welten für das Internet. Addison-Wesley

[9] Junker G (2006) Pro OGRE 3D Programming, Apress

[10] Repplinger M, Löffler A, Rubinstein D, Slusallek P(2008) URay: A Flexible Framework for Distributed Rendering and Display. Technical Report 2008–01, Informatik, Universität des Saarlandes

[11] Lohse M, Winter F, Repplinger M, Slusallek P (2008) Network-Integrated Multimedia Middleware (NMM). In: Proceedings of ACM Multimedia, 2008

[12] Bues M, Gleue T, Blach R (2008) Lightning - Dataflow in Motion. In: Proceedings of the IEEE VR 2008 workshop "SEARIS-Software Engineering and Architectures for Interactive Systems", 2008

[13] Rubinstein D (2005) RTSG: Design and Implementation of a Scene Graph Library based on Real-Time Ray Tracing, diploma thesis, Saarland University

Author

Philipp Slusallek is Scientific Director at the German Research Center for Artificial Intelligence (DFKI), where he leads the research area "Agents and Simulated Reality" since 2008. Since 1999 he is also full professor for Computer Graphics at Saarland University, Germany, where he is Principal Investigator in the large federal Excellence Cluster "Multimodal Computing and Interaction" and serves as founding speaker of the

Competence Center for Computer Science of the University. Before joining Saarland University he was visiting assistant professor at Stanford University Graphics Lab. Philipp Slusallek has published numerous scientific papers in the area of computer graphics, digital media, virtual reality, and related fields. He received a Diplom (M.Sc.) in Physics from Tübingen University and a Dr.-Ing. (PhD) from the University of Erlangen, Germany.

Corresponding author

Philipp Slusallek(✉)
German Research Center for Artificial Intelligence (DFKI)
Stuhlsatzenhausweg 3, Campus D3.2, 66123 Saarbrücken, Germany
E-mail: philipp.slusallek@dfki.de

Instantreality — A Framework for Industrial Augmented and Virtual Reality Applications

Johannes Behr[1], Ulrich Bockholt[1] and Dieter Fellner[1,2]

1. Fraunhofer-Institut für Graphische Datenverarbeitung
2. Interactive Graphics Systems Group (GRIS), Department of Computer Science, Technisch Universität Datmstadt

Abstract

Rapid development in processing power, graphic cards and mobile computers open up a wide domain for Mixed Reality applications. Thereby the Mixed Reality continuum covers the complete spectrum from Virtual Reality using immersive projection technology to Augmented Reality using mobile systems like Smartphones and UMPCs. At the Fraunhofer IGD the Mixed Reality Framework instantreality (www.instantreality.org) has been developed as a single and consistent interface for AR/VR developers. This framework provides a comprehensive set of features to support classic Virtual Reality (VR) as well as mobile Augmented Reality (AR). The goal is to provide a very simple application interface which includes the latest research results in the fields of high-realistic rendering, 3D user interaction and total-immersive display technology. The system design is based on various industry standards to facilitate application development and deployment.

Keywords

X3D, Semantic Modeling, Computer Vision based Tracking, Mobile Systems

1 Instantreality VR System

The instantreality framework is a high-performance Mixed-Reality (MR) framework that provides a comprehensive set of features to support classic Virtual-Reality (VR) and advanced Augmented Reality (AR) equally well. In this context, the term "mixed reality" describes the fusion of the virtual and real worlds as follows:

- Within Virtual Reality scenarios the user interacts with 3D digital objects in

Fig. 1 Immersive experience of machine construction with instantreality

real time (see Fig. 1). Using stereoscopic projection systems and multimodal interaction devices, the user is completely immersed in a virtual world.

- Augmented Reality (AR) describes the real-time overlay of our real environment with digital information. A typical Augmented Reality System includes a mobile computing unit, a video camera —which captures the user's environment — and a head mounted display for visualising the superimposed digital information.

The instantreality framework has been developed in close cooperation with the industry and supports various ISO/ECMA standards. It includes the latest research results in the fields of high-realistic rendering, 3D user interaction and total-immersive display technology. In Virtual Reality particular instant reality addresses the following topics:

1.1 Clustering and Distributed Rendering

In Darmstadt instantreality drives the high-immersive stereo projections system HEyeWall 2.0. This tiled display integrates 24 (6 × 4) DLP stereo projection systems and thus offers a resolution of 8400 × 4200 pixels (see Fig. 2). The system uses 48 PCs for parallelized rendering. Using colour calibration and geometric adjustment

Fig. 2 HEyeWall 2.0 — tiled stereoscopic Virtual Reality display

homogeneous and edgeless pictures are rendered. The modular system can be scaled arbitrarily. Consequently, a wide range of projection resolutions and shapes is possible. With this technology, our customers are able to visualize their 3D models, sketches, and processes in real-time on the HEyeWall.

Thereby, the instant**reality** software framework includes various optimisation methods to exploit all available hardware resources and to reach global application specific runtime goals:

- Cluster-based Rendering
  Different sort-first/sort-last algorithms balance the rendering load for every cluster-node in real-time [1]. The method scales almost linear and is not fixed to the number of CPUs/GPUs. The algorithm allows increasing the overall cluster performance by increasing the number of render-nodes.
- Multi-Resolution Datasets
  Instant**reality** can create and manage multi-resolution datasets for points, meshes and volume datasets. This allows the system to control the overall render performance to reach global application goals like a minimal frame rate.

1.2 Semantic Modelling

With the Geometric Modelling Language (GML) developed at the Institute of Computer Graphics and Knowledge Visualization at Graz University of Technology high-level semantics have been included into the modelling process [2]. Thus, 3D-Modells are not only defined with geometric primitives like points and triangles but sequences of operators and parameters are used to describe structures and repetitive forms. Using this technology different instances of a generic 3D-Modell class can be generated by just modifying specific parameters (see Fig. 3). The GML has been included into the instant**reality** framework and this yields a high potential for interactive design applications.

Fig. 3 Modelling different instances of a wheel rim using the Geometric Modelling Language

1.3 Interactive VR with Multi-touch

Multi-touch technology is one of the most interesting research fields in today's Human-Computer-Interface area. By now, this kind of technology is just the basis for a lot of new techniques in the way we are working with computers. Thereby, Multi-touch interaction can be both, interacting with more than one or two fingers simultaneously, as well as working collaboratively in a multi-user scenario. The multi-touch technique has been integrated into instantreality and it is very promising since it allows people to interact seamlessly with what they see by simply touching it [3]. A multi-touch with a typical size of 150 × 90 cm has been developed and it is used for 3D-visualisation, navigation and interaction (see Fig. 4). In this implementation the projection unit is embedded inside the table. A wide-angle optic and a mirror system create a clear and high-resolution image on the table's surface. An optical touch-sensing method tracks the user's fingers using computer vision techniques. The multi-touch table's acrylic sheet is illuminated by infrared light. As the refractive index of acrylic is higher than the refractive index of air, light does not escape but is subject of total internal reflection. Whenever an object comes close enough to the surface, the total reflection is hindered; light therefore dissipates and illuminates the object. This would also illuminate a fingertip touching the surface and a camera trained on the table's surface can now capture the resulting light blobs from fingers. The blob detection processes the recognition of bright spots in the image, which results in a set of 2D images showing fingertip positions. Blob tracking assigns a unique ID to each blob and tracks this blob from frame to frame.

Fig. 4 Interaction with virtual world using the multi-touch-table

2 Instantreality AR system

Computer-supported information and communication systems play a decisive role in shaping our work and leisure time. Not only can new technologies for human-machine interaction be found in the workplace, they are also opening up numerous new fields of application. Augmented Reality plays a major role in this context. With this technology, digital information can be superimposed upon our real environment. The AR systems are characterized by the following characteristics:

- **Mobility:** Augmented Reality technology is particularly well-suited for mobile applications. For example, it can be used to support the targeted work of an assembler in a large factory hall (see Fig. 5) or to realize an information system for tourists to be used outdoors.
- **Real-time capability:** The digital information is always superimposed on the real objects in real time — that is, the Augmented Reality system supports the continual alignment of virtual and real objects.

- **Relation to context:** Information is visualized through superimposition on real objects. This ensures that the relationship between computer-generated objects and the real 3D scene is clear.
- **Intuitiveness:** The objects which are displayed consist primarily of graphic animations which guarantee that information is transmitted in a way that is easy to understand and independent of language.
- **Interactivity:** Innovative interaction paradigms which go beyond mouse/keyboard entry support the necessary mobility (the user has to be able to move freely over a large area) as well as agility (the user must be able to act freely with both hands).

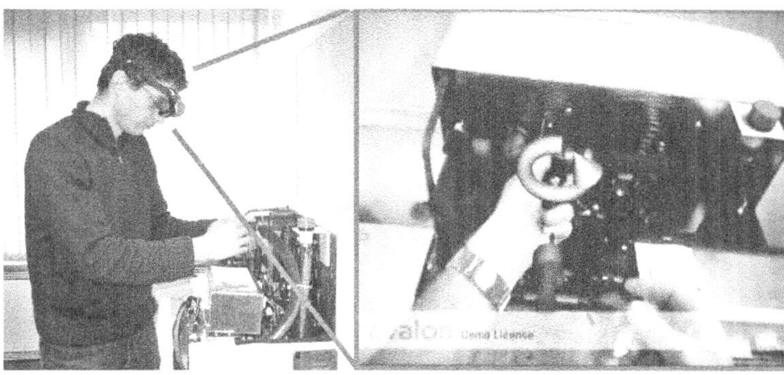

Fig. 5 Augmented Reality for the support of a service technician

2.1 Robust and Markerless Tracking

A special challenge in the development of Augmented Reality systems lies in the realization of adequate and robust tracking methods. The tracking system is used to identify and register the exact position and orientation of the user. In order to create such a solution, a hybrid, markerless tracking method has been developed that is based on the following data sources:

- **Live video data streams:** The live video data is recorded with a miniature firewire camera. This camera takes color pictures with a resolution of 640×480 pixels at a frequency of 30 Hz.
- **Inertial sensors:** The inertial sensors register accelerations in a translatory direction and velocities in a gyratory direction. Orientation in the earth's magnetic field is also registered (compass sensor). The data is exported at a very high frequency of around 100 Hz via a USB interface. The camera and sensors are combined in a tracking box, and the electronics are merged. The recorded data is correlated with the video grabber (interface in the computer for reading camera images), creating a synchronized data stream of video images, linear accelerations, angular velocities and orientation in the earth's magnetic field which serves as input data for the tracking software.

The tracking system is realized algorithmically by imitating the human sense of orientation [4]. Priority is given to processing the visual information, though in cases of bad lighting or high acceleration, the inertial systems come to the fore. For processing the visual information, the live images from the video camera are processed in real time. Features are extracted from the images which make it possible to identify characteristic features in the real environment (landmarks). These landmarks are identified by using point and edge detectors (see Fig. 6).

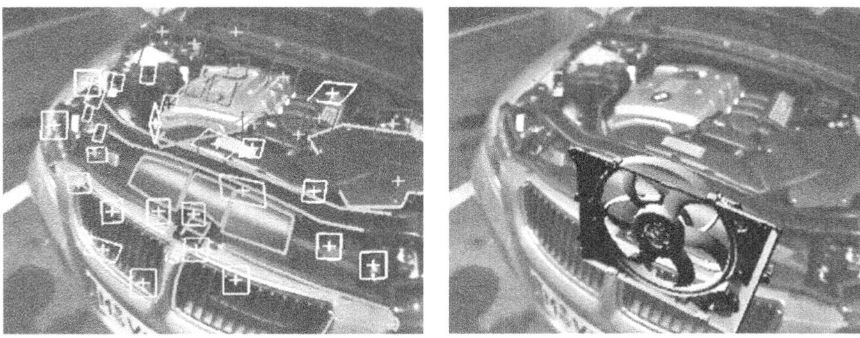

Fig. 6 Markerless tracking using natural feature tracking (left: Feature point and edges, right: AR visualisation)

With the help of the landmarks that have been identified in the two-dimensional camera images, the position of the camera is reconstructed in three-dimensional space. Epipolar geometry techniques are used to do this. Landmarks are either set in relation to previous video images (frame-to-frame tracking), or they are correlated with a digital 3D model of the real environment. The digital 3D model of the environment is created in a preprocessing stage on the basis of 3D CAD models of buildings or machines, for example. Inertial sensors support this image-based tracking method. The assumed pose (position and orientation) of the tracking unit can be predicted by analyzing the measured motion path of the tracking unit. This estimated pose is the starting point for the image-based tracking algorithms. The inertial sensors are also evaluated if no landmarks are detected in the camera images. The problem with using inertial sensors, however, is that measurement errors can accumulate, since they are not absolute poses which are registered here, but rather relative movements. For this reason, the poses determined using the inertial sensors are continually corrected with the help of the absolute pose results from the image-based tracking.

2.2 Augmented Reality Embedded Systems

Virtual Reality is often linked to technically most demanding installations whereas the intended platforms for Augmented Reality are mobile devices like PDAs or UMPCs (see Fig. 7). So, downscaled Augmented Reality systems have been developed that support different lightweight devices even on embedded operating systems. Consequently, the rendering component is based on the OpenGL ES standard. OpenGL ES supports the display of three-dimensional computer graphics on mobile end devices [5]. Building on this, a scene graph is developed with which the 3D objects can be structured and managed. The scene graph is compatible with the X3D standard.

Fig. 7 AR-maintenance support system running on the PDA

3 Conclusion

Instantreality covers the complete domain from Virtual Reality to Augmented Reality applications. Thus, Computer Vision based tracking can be used for VR applications or VR simulations can be combined with real captured environments, to mention only two examples. This fusion of CV and CG includes the simulation of consistent illumination in Augmented Reality applications as well as colorimetric measurements to support colour-confident rendering. The instant**reality** framework provides a complete set of tools and plug-ins to ease the integration of different data sources as well as the application development and deployment:

- **Integration:** Plug-ins for the most common digital content creation tools (e.g. Maya, 3DMax) enable the application developer to integrate 3D data very efficiently. Data Importers for the Framework can directly read and process the most common CAD data formats (JTOpen, Catia5, Catia4, Step, STL etc.).
- **Composition:** A special runtime environments allow the developer to integrate and compose data from different sources. The system includes various plug-ins to enable the developer to create any type of application logic and behaviour by defining components, component relations and component processing units. A nifty event and script debugger ease the development process.
- **Deployment:** Various server and middleware systems can be utilized to deploy the final applications on a wide number of hardware platforms. The server and services communicate using standard ZeroConf mechanisms to ease the installation and service process.

The system design includes various industry standards to ease the development and application service process [6]: OpenGL 2.0 (Khronos Group), GLSL (Khronos Group), Collada (Khronos Group), CG (NVIDIA corporation), X3D (ISO/IEC 19775:2004), ECMAScript (ISO/IEC 16262:2002), JAVA (Sun corporation), SOAP (W3D SOAP V1.2), ZEROCONF (IETF Zeroconf WG). All system components are available on a wide number of hardware and software platforms: Windows 2000/XP/Vista, Windows CE, Linux32, Linux64, Mac OS X, IRIX, SunOS.

References

[1] Roth M, Rieß P, Reiners D (2006) Load Balancing on Cluster-Based Multi Projector Display Systems. In: Joaquim Jorge, Vaclav Skala (eds) Proceedings of the International Conference in Central Europe on Computer Graphics WSCG, Plzen

[2] Berndt R, Havemann S, Settgast V, Fellner D (2008) Sustainable Markup and Annotation of 3D Geometry. In: Ioannides, Marinos et al (eds) Proceedings of the 14th International Conference on Virtual Systems and Multimedia, Limassol

[3] Jung Y, Keil J, Behr J, Webel S, Zöllner M, Engelke T, Wuest H, Becker M (2008) Adapting X3D for Multi-touch Environments. In: Spencer, Stephen N. (eds) Proceedings of the 13th International Symposium on 3D Web Technology, Los Angeles

[4] Bleser G, Stricker D (2009) Advanced Tracking through Efficient Image Processing and Visual-inertial Sensor Fusion. Computers & Graphics 33(1): 59–72

[5] Rieß P, Stricker D (2006) AR On-demand: A Practicable Solution for Augmented Reality on Low-end Handheld Devices In: Müller S, Zachmann G(eds) Proceedings of the 3. Workshop der GI-Fachgruppe VR/AR, Aachen

[6] Behr J, Dähne P, Jung Y, Webel S (2007) Beyond the Web Browser - X3D and Immersive VR. In: Proceedings of the IEEE Virtual Reality 2007, IEEE Symposium on 3D User Interfaces, Charlotte

Authors

Dieter Fellner is a professor of computer science at the Technical University of Darmstadt, Germany, and the Director of the Fraunhofer Institute of Computer Graphics (IGD) at the same location. Previously he has held academic positions at the Graz University of Technology, Austria, the University of Technology in Braunschweig, Germany, the University of Bonn, Germany, the Memorial University of Newfoundland, Canada, and the University of Denver, Colorado. He is still affiliated with the Graz University of Technology where he chairs the Institute of Computer Graphics and Knowledge Visualization he founded in 2005. Dieter Fellner's research activities over the last years covered algorithms and software architectures to integrate modeling and rendering, efficient rendering and visualization algorithms, generative and reconstructive modeling, virtual and augmented reality, graphical aspects of internet-based multimedia information systems and digital libraries. In the latter field he has coordinated a strategic initiative funded by the German Research Foundation (Deutsche Forschungsgemeinschaft) from 1997 till 2005. In the areas of computer graphics and digital libraries Dieter Fellner is a member of the editorial boards of leading journals and a member of the program committees of many international conferences and workshops. In 1997 and 2003 he was program chair of the Eurographics conference. He is a member of EUROGRAPHICS where he serves in the Executive Committee, ACM, IEEE Computer Society, and the Gesellschaft für Informatik (GI) where he serves as a member of the extended Board of Directors and as Chairman of the Graphics Chapter (Fachbereich Graphische Datenverarbeitung). Furthermore, Dieter Fellner is an advisor for the German Scientific Council, the German Research Foundation, and the European Commission (as a member of ISTAG).

Johannes Behr leads the VR group within the department "Virtual and Augmented Reality" at the Fraunhofer-Institut für Graphische Datenverarbeitung in Darmstadt, Germany. His areas of interest focus on virtual reality, computer graphics and 3D interaction techniques. Most of the results of his recent work are available as part of the InstantReality Framework. He has an MS from the University of Wolverhampton and received his doctoral degree from the Technische Universität Darmstadt.

Ulrich Bockholt is the head of the department "Virtual and Augmented Reality" at the Fraunhofer Institute for Computer Graphics (IGD) in Darmstadt, Germany. He received his diploma degree in Mathematics from the University Mainz (Germany) in 1997 and his doctoral degree from the Darmstadt University of Technology in 2003. Since 1997 he is working at the Department "Virtual & Augmented Reality" at the Fraunhofer-IGD in Darmstadt. In 2004 he has received his doctoral degree at the Technical University Darmstadt- He participated or led several national and international research projects in the area of virtual and augmented reality, such as SKILLS, VIVERA and AVILUS.

Corresponding author

Johannes Behr(✉)
Fraunhofer-Institut für Graphische Datenverarbeitung, Fraunhofe- rstraße 5, D-64283 Darmstadt, Germany
E-mail:Johannes.Behr@igd.fraunhofer.de

Interactive Simulation Data Exploration in Virtual Environments

Marc Wolter, Thomas Beer, Philippe Cerfontaine, Bernd Hentschel and Torsten Kuhlen

Virtual Reality Group, RWTH Aachen University, Germany

Abstract

In the last few years, simulation of technical and physical processes has become an important pillar in engineering. Computational Engineering Science (CES) is nowadays an indispensable and essential tool for the development of, e.g., airplanes, cars, combustion engines, turbines etc. In this work, we present a framework that tackles the growing complexity of these simulations. For simulations of highly complex processes, we enable interactive exploration in 3D space by application of Virtual Reality (VR) technology. For simulation data that is only a part of a complete process simulation, we propose methods that facilitate data analysis within the global context of an interdisciplinary project. By considering a single simulation as well as its part in a larger context, this framework enables engineers to explore their data on multiple scales of complexity.

Keywords

Virtual Reality-based Scientific Visualization, Process Simulation

1 Introduction

In the last few years, simulation of technical and physical processes has become an important pillar in engineering. The relatively new field of Computational Engineering Science (CES) is nowadays an essential tool in basic research as well as for the development of, e.g., airplanes, cars, combustion engines, turbines etc.

In the latter application, numeric simulation can be used to improve the quality of products, provide better services around products, shorten product development time and save development costs, as costly physical prototypes and experiments can be cut down.

Even in medicine, CES is going to play an important role in the analysis of flow within blood vessels and the development of artificial blood pumps, or in order to understand the air flow within the human respiratory organs. Due to the considerable cost of experiments, flow simulations are continuously pushing forward.

The enormous growth of computing power accelerated the development of CES. Today, physical phenomena are more and more simulated in three instead of only two dimensions, based on very fine grids containing up to several million cells. In addition, researchers are investigating more and more unsteady flow phenomena, where the flow field changes over time, resulting in huge datasets especially when used in combination with highly refined grids. Due to the high complexity of today's simulated phenomena, the analysis procedure of datasets gets a more explorative character.

In contrast to a confirmative analysis, in an explorative analysis the hypotheses about the characteristics of a simulated phenomenon still have to be generated during the analysis procedure, resulting in a trial-and-error process, i.e., the engineer continuously defines parameter values to extract features which are thereafter often rejected because of unsatisfying results. Then, the parameters are modified for a renewed feature extraction. This iterative scheme is applied until a comprehension of the simulated characteristics is attained. Thus, an explorative analysis relies heavily on the interactivity of the underlying system. All in all, researchers are going to examine physical phenomena of such a high complexity that traditional methods of post-processing, like producing static images or at best animations, are neither an effective nor an efficient approach to understanding the simulated flow fields anymore. Instead, engineers demand interactive exploration of their data in 3D space, which eventually leads to the use of Virtual Reality (VR) technology.

As real-time interaction is not only a fundamental and imperative feature of VR but basically the main reason for its use, additional real-time constraints arise, which are not present in conventional, desktop-based visualization systems. The primary requirement is to persistently maintain a minimum frame rate of 30 frames per second. In addition, response time to user input should not exceed 100 milliseconds. While the latter is typically impossible to maintain with today's datasets with a size of multiple gigabytes, response times still have to be reduced as much as possible. However, staying above the minimum frame rate at all times is mandatory. Upon violation of these rules, the user acceptance decreases and an explorative analysis might be compromised. Thus, VR visualization software has to be fundamentally designed to meet the requirements of an interactive exploration.

To further aggravate the complexity of simulation data visualization, an individual simulation describes a certain phenomenon from the viewpoint of a single research field only. In order to gain a holistic view, multiple simulations of different aspects—each from a different field of expertise—need to be integrated into a single workflow. These simulations can depend on one another, resulting in a simulation chain. As an example, a single engineering process can be simulated on different scales, e.g., a microstructure simulation for a gear wheel embedded in a large-scale welding simulation.

To this end, as an additional requirement, multiple visualization results from cooperating engineers need to be seamlessly integrated in the scientific visualization process of each single user.

In order to adhere to the aforementioned requirements, we have developed the flow visualization framework ViSTA FlowLib [1] and the parallelization system Viracocha [2]. Both tools have been targeted from the very beginning at an interactive exploration of flow phenomena in virtual environments, and have successfully been used in several research projects, including industrial research.

This work is structured as follows. In Section 2, we regard the analysis of a single simulation, only. After briefly introducing the hybrid visualization environment we propose for explorative flow analysis in VR (see Section 2.1), we will describe two interactive visualization techniques building on this system (see Sections 2.2 and 2.3). These techniques serve as an example for how interactivity demands for virtual environments can be fulfilled even for data set sizes exceeding the capacity of single workstations.

In Section 3, we place this interactive exploration in the global simulation chain and describe, how multiple interweaved simulations can be explored within a virtual environment. This includes combination of single simulations using a distributed simulation platform (see Section 3.1), a necessary meta-description of heterogeneous simulation data (see Section 3.2), and certain aspects of interactivity and interaction when visualizing multiple simulations simultaneously (see Sections 3.3, 3.4 and 3.5). Section 3 describes work in progress, which is partially actively used by engineers and which partially discusses future work.

2 VR-based Simulation Data Exploration

2.1 A Hybrid Visualization Environment

In this section, we briefly describe the hybrid visualization environment we propose for explorative analysis of simulation data in VR. This is the architectural concept underlying the development of the ViSTA FlowLib and Viracocha frameworks.

In order to cope with the high demands to provide interactivity for VR-based simulation data exploration, we propose to use a system comprising multiple components specialized for certain subtasks. To speed up the exploration process, all available resources have to be employed in a sensible way.

The uppermost decomposition is the separation of a visualization front-end (i.e., a dedicated graphics workstation) for graphical tasks and powerful HPC equipment for compute- and memory-intensive tasks. The HPC back-end itself connects to some file system, allowing for the access to persistently stored simulation data. In the case of a system driving an immersive virtual environment, the front-end itself is potentially a distributed rendering cluster. With the advent of multi-core processors and programmable graphics hardware, each node of the back-end as well as the front-end can be considered as containing another small-scale parallel computational resource in its own right.

In order to make optimal use of such a hybrid (and highly heterogeneous) visualization environment, the capabilities and limitations of its components have to be considered. This includes computing power, programming model, memory/storage space, and memory bandwidth.

The visualization front-end requires software components which deal with efficient rendering of data, user interaction and visualization algorithms for techniques which require low latency. The latter includes GPU-based algorithms, which are massively parallelized, but restricted by the memory capacity of today's graphics cards. Efficient rendering methods have to take the special requirements of an immersive scene into consideration, e.g. sufficient depth cues and high frame rates. The ViSTA FlowLib framework provides such software components; one example is interactive particle tracing.

Instead of providing a parallel algorithm for a specific visualization algorithm, the Viracocha back-end [2] provides the resource management and transport flow necessary for the connection of remote parallel algorithms with virtual environments. The internal structure and connection to the visualization front-end is depicted in Fig. 1a. Resource allocation includes the gathering and managing of available processing resources as well as memory resources during an exploration session.

Tasks are distributed among worker nodes— using the Message Passing Interface (MPI) for interprocess communication —which enables the application of distributed memory architectures. Scheduling schemes are chosen dependent on the exploration process (see Section 2.1). On each worker node, pipelining and thread-level parallelization are applied. Pipelining subtasks (loading, computation, and transmission) prevents tasks from waiting for I/O bottlenecks. Especially file access is a common problem for large data sets. Additional data parallel computation using a shared memory model (e.g., with OpenMP or Posix threads) allows for fast computation on each node. With different levels of parallelism provided by the system, an optimal configuration adapted to the available architecture and dataset is rendered possible.

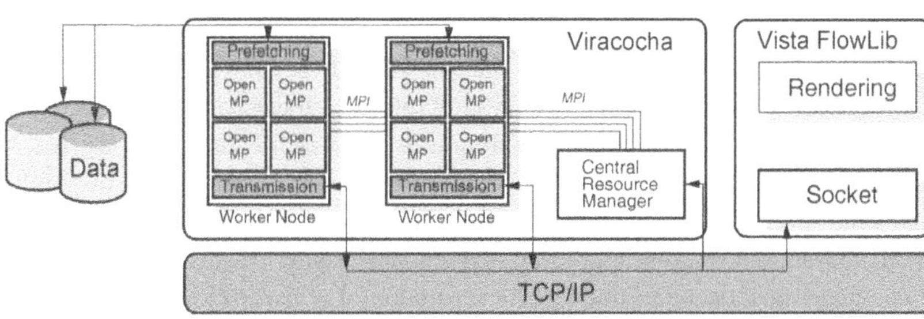

Fig. 1a ViSTA FlowLib is connected to the parallel post-processing back-end Viracocha, which typically runs on an HPC system

One of the most restricted resources is memory. Dealing with large datasets, I/O is a bottleneck for most applications. Parallel computers provide more resources in form of memory than workstations. But even when sufficient main memory is available, moving data from secondary storage can take tremendous time. Many post-processing calculations spend a good deal of their runtime loading desired data, which would be unnecessary, if an efficient data management for CFD-data sets were in use.

Therefore, besides existing techniques for more efficient secondary storage access, we include caching of data for later re-use as well as prefetching techniques for predicting disk access. As an example, data for multiple time steps is typically processed consecutively, which makes time step data easily predictable[3]. The Viracocha framework includes a data management system providing several caching and prefetching strategies.

In the following sections, we address different scheduling strategies provided by Viracocha before we introduce two interactive visualization techniques which use the described features of Viracocha.

2.2 Scheduling Techniques

Viracocha includes a set of different scheduling techniques, which distribute discrete time steps of a time-varying simulation data set to free worker nodes. The existence of Viracocha's data management affects the total runtimes obtained by different scheduling strategies.

Static scheduling strategies — that use an a-priori distribution of tasks to processes — are often avoided because they can lead to significant load imbalances. However, if data is cached at each process, using the same distribution pattern in consecutive runs results in optimal cache usage (see Fig. 1b).

Dynamic scheduling distribute tasks during runtime in order to avoid load imbalances, but thereby assign in consecutive runs the same task to different processes, which leads to inferior cache usage (see Fig. 1b).

In order to combine both advantages, Viracocha uses a scheduling strategy called *dynamic cache*. This strategy works similar to dynamic scheduling, but includes knowledge about data locality. That is, each time a worker node needs a new task, the scheduling strategies tries to assign a time step that already resides in this worker's cache. Using this approach, the number of cache hits can be significantly increased (see Fig. 1b).

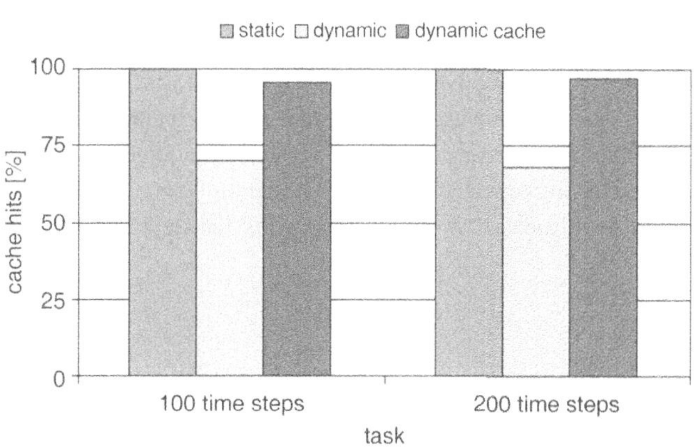

Fig. 1b Cache hits for applying different scheduling strategies (static, dynamic, dynamic cache) to visualizations tasks with 100 or 250 time steps

The chosen scheduling strategy also affects the visualization perceived by the user. The reason is that all result data, which is concurrently produced by Viracocha, is directly transmitted to and displayed by ViSTA FlowLib. If this data is observed in an animated loop, the arrival order of results influences the animation's content.

As an example, Fig. 1c depicts the arrival distribution of 5,000 discrete time steps when using a static scheduling. Darker areas arrive earlier than brighter areas. The set of time steps is distributed into eight blocks, each of which is computed by a single worker from left to right. Consequently, the animation consists of eight separated "scenes" that are continuously extended. Moreover, this distribution is independent of any user preference.

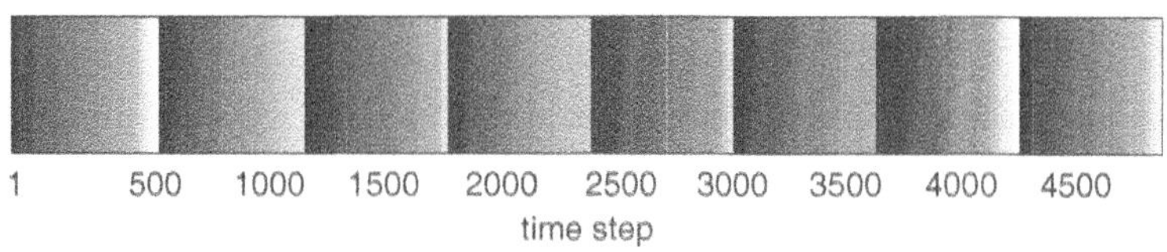

Fig. 1c Arrival distribution of discrete time step result data for a static scheduling technique. Darker areas arrive earlier than brighter areas

Therefore, we propose to adapt the scheduling according to the user's analysis goal. The basic idea is that by influencing the arrival order of results in the animation the user can decide faster if the chosen visualization suits his needs. In order to adapt the scheduling to changing user goals, we apply a dynamic scheduling strategy that in addition to the data management information as described above includes a prioritization of discrete time steps based on the user goal.

In this section, we briefly describe scheduling techniques for two common analysis goals: *obtaining an overview* and *continuous observation.*

The *overview* scheduling strategy repeatedly samples the set of time steps with an increasing sample rate. Figure 1d shows an arrival distribution achieved with this strategy. The darker regions, which arrive earlier, are equally distributed along the time range. In the animation, this results in a coarse overview of the time-varying phenomenon, which is continuously refined as the sample rate increases. This strategy is especially beneficial when the user tries to find a suitable parameter to depict a certain phenomena. By obtaining a coarse overview first, he can quickly reject unsatisfactory parameters without having to watch the whole animation first.

The *continuous observation* strategy supports the detailed analysis of a time-

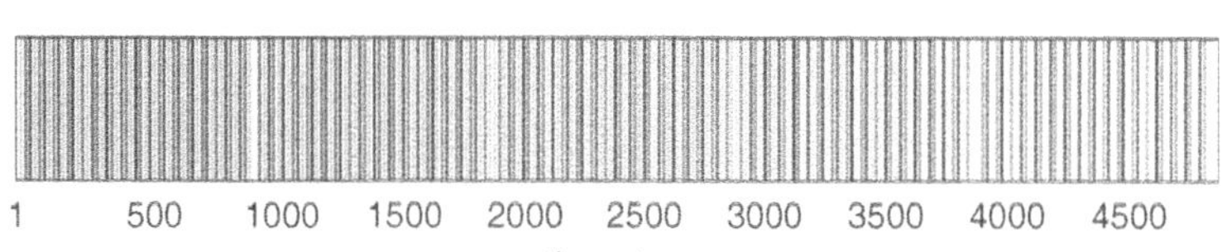

Fig. 1d Arrival distribution of discrete time step result data for the overview scheduling technique. Darker areas arrive earlier than brighter areas

varying phenomenon's evolution. By synchronization of the animation's speed and the concurrent computation using queuing models each computed result data arrives just-in-time at the animation. This results in a continuous animation without gaps, such that the user can observe the shown evolution without annoying interruptions. Figure 1e depicts an exemplary arrival distribution for the continuous observation strategy. Starting from the user's current position in the animation, results are computed in the animation's direction (i.e., from left to right). The synchronization tries to make sure that no visible gaps occur in the animation (not visible in the figure).

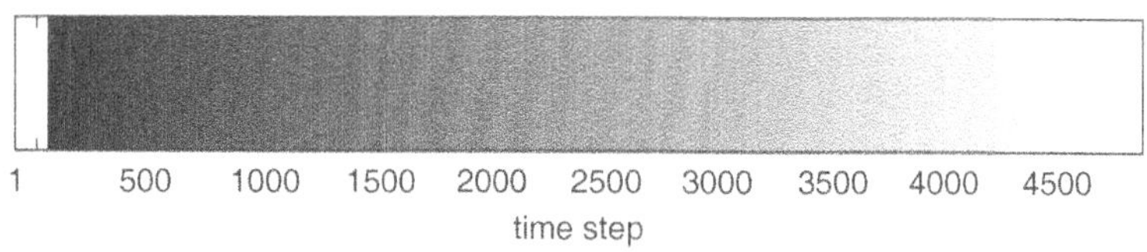

Fig. 1e Arrival distribution of discrete time step result data for a continuous observation scheduling technique. Darker areas arrive earlier than brighter areas

These two scheduling strategies are only examples for common user goals. Other strategies for problem-specific user goals are possible.

In the following, we present two visualization techniques that utilize the parallel Viracocha system.

2.3 Interactive Particle Tracing

One particularly active field of CES is the simulation of flow phenomena (also known as Computational Fluid Dynamics (CFD)).

The benefits of VR-based flow exploration become apparent most clearly when employing visualization methodology with immediate feedback. An example is given by interactive particle tracing, where the user can directly inject particles into the dataset, which are then immediately advected by the flow field.

Utilizing large numbers of such particles facilitates an intuitive understanding of the underlying flow features. Navigation and particle injection using a spatial 3D input device inside a fully three-dimensional environment allow a natural and intuitive exploration of the data at hand and is clearly superior to the often cumbersome interaction as performed on desktop visualization systems, which typically leads to an inherent discrepancy between a complex three-dimensional problem domain and a visualization system with merely two-dimensional input and output.

To cope with the computational load of this process, pre-computation can be used to avoid extensive calculations during the interactive exploration phase [4]. However, this approach relies on a time-consuming preparation phase and significantly increased storage requirements.

Another solution is to change the underlying process. As users often concentrate their analysis on a specific spatial region, we chose to integrate the definition of a region-of-interest explicitly into the analysis process [5] (see Fig. 2).

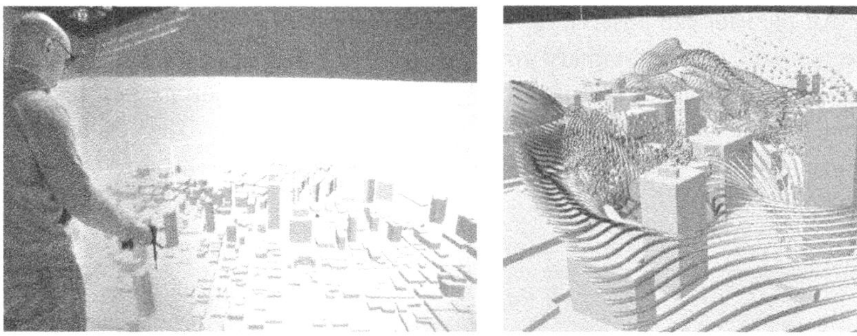

Fig. 2 Analysis of a user-defined region-of-interest. The region is defined in the virtual environment (left) and then analyzed, for instance, via particle tracing (right)

The user selects a region-of-interest inside the virtual environment, for which data of a defined size is prepared on the HPC back-end. Resampling of an unstructured grid to a computationally more convenient uniform grid can be efficiently parallelized with OpenMP, while time-varying data is distributed to different processing nodes using MPI, which leads to a waiting time of a few seconds for reasonable data sizes.

This data is then sent to the visualization front-end, where it is interactively explored, for instance using GPU-based particle tracing. The reduced data size is chosen such that the time-varying region-of-interest fits into graphics memory.

Inside the region-of-interest, the user can position arbitrary seeders, which are graphical widgets injecting massless particles into the flow using a defined pattern. The advection of these particles is done on the GPU, based on the reduced data. Combining tracing and rendering of the particle traces on the same subsystem avoids additional latency. For an efficient rendering of these traces, either as particles or as pathlines including the history trajectory, billboarding techniques are applied [6].

As the data preparation typically involves lossy data reduction, the introduced error is conveyed to the user in order to allow for an evaluation of the precision of the visualization, for instance using volume rendering. In addition, on the reduced data, arbitrary visualization algorithms can be executed to support the free exploration of the region-of-interest [7] (see Fig. 2).

2.4 Interactive Feature Definition

Feature-based visualization tries to tackle the large-data problem by automatically extracting predefined structures of interest — the so-called features — from the data. Classic examples for features from the field of fluid mechanics include vortices and shocks.

However, many features are not clearly definable in a binary fashion suitable for automatic extraction. We propose that this problem can be tackled by a human-in-the-loop approach to feature definition and analysis. Therefore, our goal is to establish a method that allows the interactive specification of arbitrary features.

In general, our visualization approach relies on multiple linked views. This is a concept originally developed in the field of information visualization, where changes to on view are immediately propagated to all linked views. This concept is combined with brushing, which enables the user to interactively highlight and manipulate a subset of data items from a larger set. We have concentrated on the use of three-dimensional scatterplots which are presented and interacted with in a virtual environment, as shown in Fig. 3. Each plot shows an arbitrary combination of three data attributes which are recorded per data point. In the case of simulation data analysis, this allows the domain expert to assess relationships between a multitudes of variables. In addition to simulated variables, the spatial positions of the data points may be used for display, resulting in a three-dimensional representation of the original simulation domain.

Brushing is used to highlight a set of selected points by marking them with a three-dimensional, rectangular selection box. In effect, this box is a manipulability visual representation of three one-dimensional range queries. The box can be created inside the data by direct interaction. To enable an iterative refinement of the selection query, the user can modify each side of the box individually.

Each query defines an inequality of the form $a \leqslant x \leqslant b$. Brushing now highlights all the data points whose data values satisfy all the inequalities. In order to decrease the amount of generated information, these data points are typically reduced using a binning approach, that is, data points are accumulated in predefined regular bins. This is useful, as single data points are not of interest to the domain scientist, only a representation of the spatial distribution at a sufficient resolution. The resulting selection is then displayed synchronously in all plots using color coding.

The fundamental idea of this brushing and linking-based approach is to quickly generate and check hypotheses about features in the data based on their attribute values. This is done by marking interesting data ranges, which might characterize the feature in question. This is done in plots, which show data attributes, e.g., pressure or temperature at a certain location as illustrated in Fig. 3. Then the user cross-references the selection to spatial plots in order to find out, where exactly the marked values occur in the data set. In this way, the definition of features is interactively specified by the domain expert rather than being contained in a "black-box" algorithm.

Fig. 3 A user interactively marks interesting value ranges in three-dimensional scatterplots. The combination of direct interaction and quantitative data display allows domain experts to control the feature extraction process

However, in order to evaluate these queries, the whole data set has to be searched. Using indexing techniques is only useful to a certain degree, as the selectivity of the combined queries can be arbitrarily low. Therefore, this subtask is moved to the HPC back-end. Time-varying data is distributed to different compute nodes, while the evaluation of the queries is parallelized using OpenMP. Selected data of the active plot is shown immediately, the linked plots are updated when the results from the back-end are available. While the number of MPI nodes reduces the total execution time, the application of OpenMP significantly reduces the waiting time until the first result is ready [8].

During an exploration session, the queries will frequently be changed or refined until a sufficient understanding of the data has been established. The hypotheses, which result from the exploration phase, may later be tested in detail with dedicated visualization setups. Therefore, the method proposed here is meant to enhance rather than to replace standard visualization approaches. Initial evaluations conducted in close collaborations with domain experts showed promising results.

3 VR-based Process Exploration

In our projects, immersive VR has not only proven to be a helpful tool for experts, but has also shown its potential in joint projects that cross disciplinary boundaries. Being close to an exploration in real wind tunnels, in particular real-time particle tracing has turned out to be an appropriate method to intelligibly communicate characteristics of even complex, unsteady flow phenomena to non-experts (see Fig. 4 left).

Fig. 4 Immersive investigation of fluid flow: Efficient depiction of particle trajectories via bill-boarding inside a blood pump (left) and volume rendering of temperature within a film cooling simulation (right)

Interactive definition of features has shown to be a useful tool for a more quantitative analysis of correlations between multiple data attributes, which especially target the needs of simulation experts (see Fig. 4 right).

Both techniques are only useful when waiting times for the user can be reduced to an amount acceptable in virtual environments. This is made possible by a hy-brid visualization environment, where specific subtasks are distributed to the most suitable component.

Once the user has explored his/her simulation and gained a certain amount of insight about the phenomenon under investigation, it might still be necessary to consider additional information sources to fully understand the representation of the simulation results. This is especially true in interdisciplinary projects where experts from different research fields are simulating separate aspects of the same phenomenon to contribute to more precise predictions.

A promising technique to increase accuracy in this case seems to be linking the simulation programs by making them exchange simulation results (see [9] and Fig. 5), e.g., to replace general assumptions in subsequent simulations by specifically computed information from prior simulations.

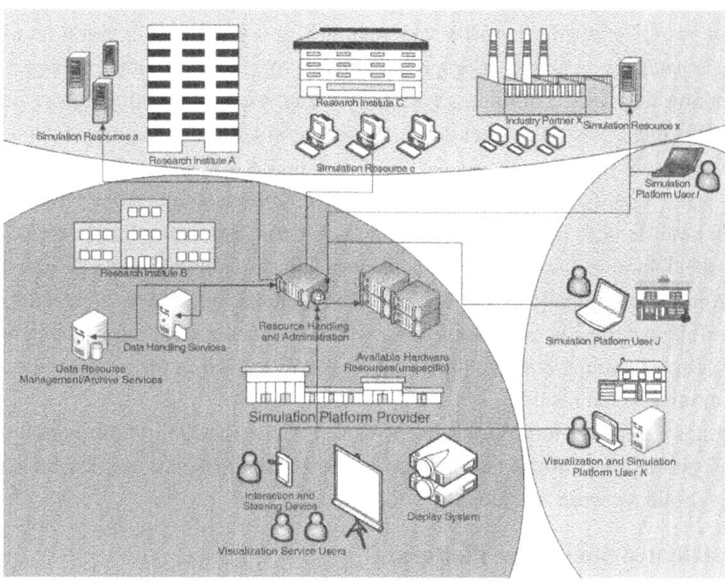

Fig. 5 A simulation platform that allows multiple users to take advantage of various, distributed simulation resources

This rather peculiar context of having multiple simulation data sets of various types concerning a common topic allows a more comprehensive representation of the subject that is being investigated. Since traditional visualization solutions usually don't provide the functionality required to investigate multiple data sets other than one at a time, we are currently enhancing the aforementioned ViSTA FlowLib toolkit [1] to support the simultaneous analysis of related simulation results in virtual environments.

While comparison is certainly an important aspect of understanding the interrelationship between simulations, this is not to be mistaken for comparative visualization where variants of the same simulation run or alternative solutions have to be evaluated side by side. The main difference comes from the fact that the relation between the data being compared is far from obvious because of its heterogeneity and the complexity of the underlying simulation chain.

The more interweaved the simulation programs become, the more complicated it gets to analyze the relationships and the impact on the final simulation result. This analysis can only be accomplished involving domain experts that are familiar with not only their specific area of expertise but also with the neighbouring disciplines. Depending on the number of simulations employed in the modelling process, several experts might be required to explain and understand the simulation results.

Immersive virtual environments, in particular room-mounted systems, provide the benefit that several scientists can easily work together. Although they are usually designed to provide ideal vision for a single viewer, depending on the scenario being displayed they allow multiple users to get a common grasp of the data [10].

We propose to split up the analysis process into an in-depth analysis of the specific simulation results of each scientist at first and then, in a later step, exploration of all the individual visualization results appropriately pieced together in a common context.

Breaking down the complete process of visualizing and analyzing the data into these two distinct phases may seem a rough cut but also provides some advantages that may not seem obvious at first glance. The first phase is the traditional visualization scenario where one domain expert is making sense of new results from a simulation he is already familiar with (e.g., using the techniques described in Section 2). In the second phase of the domain encompassing visualization, all available visualization and analysis results from other simulation experts have to be included into the visualization process to successfully achieve the analysis goal — which is to understand the interdependencies between coupled simulations in all their ramifications as well as the implications for and the impact on the final results.

In this section, we introduce current work on visualization of simulation chains. While the simulation platform (see Section 3.1) is already in active use, Sections 3.2–3.5 present preliminary work and ideas.

3.1 Distributed Simulation Platform

To enable data exchange between various simulation programs we suggest to take advantage of distributed *simulation resources*. A simulation resource can be any combination of hardware, operating system and software for as long as it provides simulation functionality. To successfully incorporate such a large line up of systems we heavily rely on Condor [11] to network all the resources on the platform. Condor provides enough flexibility and portability to successfully abstract from system specificities. In our context, Condor is employed for managing specific computation resources rather than distributing compute jobs to unspecific systems in a grid environment. Although this is not really the intention behind Condor, it has proven to be a reliable basis to develop a simulation service oriented architecture for simulation experts. Condor also allows chaining simulation jobs together which is the most crucial asset for integrating simulation programs. Implementing an already defined workflow for different simulation programs is tedious work but mostly feasible using the directed acyclic graph manager of Condor.

On the other side, providing the means to flexibly build chains with potentially arbitrary combinations and sequences of simulation programs requires further research.

To this end, we are currently developing a new internet based user interface to construct comprehensive simulations of production processes. To facilitate the communication between the actual simulations the common "language" on the simulation platform is the Visualization ToolKit (VTK) [12]. Every simulation providing result data for any other simulation on the platform uses the most appropriate format available in VTK. As a software library, VTK already delivers useful algorithms that help bridging gaps between simulation programs, for instance transformation from hexahedral to tetrahedral grids, or an easily available desktop visualization tool like ParaView. The most important part for the simulation platform is the fact that a common agreement on the language has been reached. This maximizes the reusability of code and also eases the joint visualization currently under development.

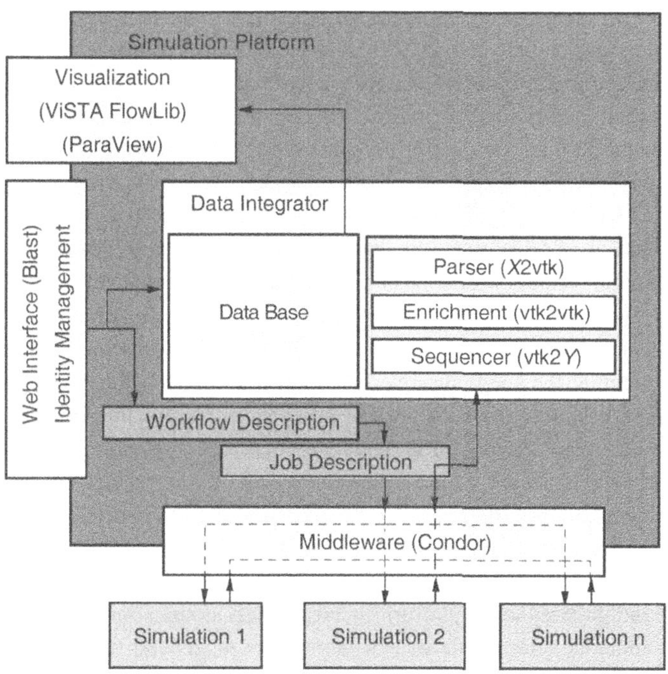

Fig. 6 Architecture of the simulation platform capable of simulating processes through chained simulation resources

3.2 Meta Data Model

The simulation platform, as described beforehand, serves as a means to gather results from different simulation tools that are chained to simulate a whole process. Each tool is used to simulate one or more different aspects of this process. Visualizing the whole process is first of all facing the task of bringing those different aspects back into their common process domain context.

More precisely, the visualizations of each data set have to be arranged in a way to fit into their corresponding spatiotemporal extents within the whole process context. For single simulation tools, the context is usually neglected (as far as the underlying simulation model does not depend on it), so that this information is lost when investigating a single data set, only.

Although we have established the formats provided by VTK (see Section 3.1) as a common language for simulation results, this does not cover things like units and ranges of the data values. In the most trivial case, all those units and ranges match across all simulation data sets. Thus it might suffice to find the correct position for each single data set within the whole process to compound a combined visualization. But already in this trivial case, the availability of meta-information is mandatory, as it is needed to classify it as a trivial case. Nevertheless, the common case requires to simulate across different scales of scope. Therefore it is not meaningful just to arrange the visualizations according to their real-world spatiotemporal positions, but more detailed information describing the relation between different simulations are necessary. As in reality, one would not see the micro-scale simulation parts when viewing macro-scale simulations and on the other hand not get the macro-scale context when zooming in for analyzing a micro-structural simulation result. For this case, more abstract methods to clearly depict relations between the data sets of different magnitude have to be found to provide meaningful views of the simulated process.

In the case of transient data sets, the simulations often have discrete time steps in different temporal resolutions that need to be matched into the overall temporal context. To achieve this, proper knowledge about the specific temporal characteristics of the simulations is mandatory. From a technical point of view, this resembles matching a list of files representing the single discrete time steps that may have any (potentially non-linear) temporal relation to the simulated overall process. Again, this kind of information is usually not provided directly in the files present on the simulation platform and thus has to be carried by means of a meta-data structure.

After these examples of what kind of meta-information is needed to enable even basic kinds of compound process visualization, the subsequent problem arises: Where can this needed meta-information be derived from?

As the whole simulation platform itself as well as the simulation tools running within are under development, an automatic way of getting the meta-information consistently through a simulated process chain is not yet feasible. As some information cannot be derived automatically from the given simulation data, it needs to be queried from the user during the process of setting up a process simulation chain. Other information like which files belong to a single data set are not determinable retrospectively, but should be journalized by the simulation platform. Although it is generally advisable to avoid mixing auto-generated and user-supplied information, we regard this a practical way to get the meta-data structure on its way in the first place.

This points out another requirement for the meta-data structures: They need to be very flexible and dynamically extendable to incorporate data that has yet to be considered during the rapidly evolving development process of the simulation platform. Even later on, when the platform itself may have settled to a stable state, the ever changing combination of new and old simulation tools may most likely require new meta-information to be incorporated.

3.3 Asserting Interactivity

Another problem resulting from combining different visualizations in a single application is a major performance issue. Even one single visualization can exhaust the available resources of a desktop system. As mentioned in Section 1, in particular when immersing the user in a virtual environment, interactive frame rates and response times need to be ensured. When simultaneously rendering several visualizations created by different scientists in a single scene, these interactivity requirements are easily violated.

To counteract this problem, for polygonal data automatic reduction techniques are employed. These reduce the polygonal visualization data to a degree that can be interactively handled by the available graphics subsystem, thus asserting a stable and interactive frame rate.

Though, this procedure affects the visible representation of the data. Depending on the degree of decimation, certain details are subject to reduction. We propose that the user takes full control over the shown level of detail. That is, the user can request the original data resolution at any time, by giving up on the frame rate warranty in the process of doing so. Using available VTK algorithms enables decimation of all involved polygonal visualization data. This is done without another conversion step, as this format has been accepted as communication language to transfer numerical results on the simulation platform (see [9] and Fig. 6 for further information on the subject).

3.4 Explorative Analysis of Multiple Data Sets

In Section 2, two interactive exploration techniques for single data sets were presented. The exploration of multiple visualizations in a common context however brings up new questions.

The presence of different transient data sets, with different spatiotemporal interrelations, opens up a high number of potentially controllable degrees of freedom. This number has to be greatly reduced to avoid the user from getting lost or even navigating into meaningless configurations and — in the worst case — misinterpret the simulation results because they are displayed in a way that confuses their relations. Thus, keeping relations between the data sets consistent is of the utmost importance in this context.

For exemplification: During the simulation of a production process several heat-treatment steps of a metal part occur. These are simulated in two scales, macro-structural simulation of heat distribution and micro-structural simulation of grain structures at characteristic points. For the explorative analysis of the visualized data sets it makes sense to compare the different behaviours during heat treatment, e.g., before and after a forming step. If the user has complete unconstrained control over all spatiotemporal degrees of freedom, he/she will likely drag the visualizations around in space and modify their temporal positions to compare the transient heating processes side-by-side.

As a result, the user may — without noticing it — possibly derive wrong interpretations from the micro-structural pre with the macro-structural post-forming heat-treatment simulation because of the different, but similarly looking visualizations. The same applies to spatial relations only if there are no outstanding visual cues that differentiate the single data sets.

Spatiotemporal relations and relations of different scales of scope between data sets have to be defined, most probably embedded into our meta-data structure. We still need to evaluate how strong the degrees of freedom should be constrained to reach a balanced level of interactivity that prevents the overall process visualization from getting too complicated — while at the same time not constraining the opportunity to freely explore all parts of the process. It has at least to be asserted that the user can at any time retrieve hints about the relations between data sets.

To conclude, handling multiple visualizations compounding the overall process at different scales of scope demands for interaction metaphors that allow to navigate in a multidimensional space characterized by spatial and temporal dimensions as well as a dimension of scale of scope. The user should be able to zoom into (e.g., micro-structural) details while still keeping the relation to a higher scale of scope context present. Finding solutions to this problem that fit the needs of immersive environments is a very challenging task we are currently rolling up to provide the different domain experts with the opportunity to gain the most insight out of their coupled simulation results.

3.5 Linking Visualizations

An open topic is how different experts exchange information that accompanies produced visualization results. Current research focuses on methods that let the users highlight and annotate areas of special interest in the simulation data [13]. The idea is to identify interesting features (either automatically or by the user's interaction) and augment such a feature with meta-data, e.g. by attaching a Post-it®-like note. Two or more of these annotations can be connected through *links*, every link possibly having its own annotation to explain the connection. This relation-based visualization combines the actual simulation results and their interpretation, meaning the possible explanation for the identified phenomena, and their interdependencies. This additional information generated by the experts needs to be made persistent to produce tangible output. Without persistence, every finding aside from the gained insight of an individual expert would vanish with the visualization application being shut down. This persistent data is incrementally extended in future sessions or used to create a report of the visualization and its conclusions.

4 Conclusion

In order to react to the growing complexity of simulation data, we have proposed the ViSTA FlowLib toolkit for interactive scientific visualization in virtual environments.

ViSTA FlowLib consists of various techniques which enable simulation data exploration on different scales.

To exploit the benefits provided by Virtual Reality technology, interactive particle tracing and feature definition have been introduced as capable techniques for an in-depth exploration of single simulation data. Interactivity requirements demanded by virtual environments are rendered possible by hybrid parallelization, which exploits multi-and many-core architectures available today.

Different methods that are necessary to analyze a simulation data set within its global context have been presented. Interdisciplinary work processes, which are getting more and more common for highly complex simulations, are supported by a shared simulation platform and visualization systems addressing the need of joint visualizations.

We believe that an integration of the techniques discussed in this paper with Virtual Reality technology to communicate information will enable interactive exploration even of future complex simulated processes.

Acknowledgments

This work has been partially funded by the German Research Foundation (DFG) as part of the Cluster of Excellence "Integrative Production Technology for High-Wage Countries".

References

[1] Schirski M, Gerndt A, Van Reimersdahl T, Kuhlen T, Adomeit P, Lang O, Pischinger S, Bischof C (2003) ViSTA FlowLib —A Framework for Interactive Visualization and Exploration of Unsteady Flows in Virtual Environments. In: Proceedings of the IPT-EGVE Symposium, pp 77 – 85

[2] Gerndt A, Hentschel B, Wolter M, Kuhlen T, Bischof C (2004) Viracocha: An Efficient Parallelization Framework for Large-Scale CFD Post-Processing in Virtual Environments. In: Proceedings of the IEEE Supercomputing, published on CD-ROM

[3] Wolter M, Hentschel B, Schirski M, Gerndt A, Kuhlen T (2006) Time Step Prioritising in Parallel Feature Extraction on Unsteady Simulation Data. In: Proceedings of Eurographics Symposium on Parallel Graphics and Visualization, pp 91 – 98

[4] Ellsworth D, Green B, Moran P (2004) Interactive Terascale Particle Visualization. In: Proceedings of IEEE Visualization, pp 353 – 360

[5] Schirski M, Bischof C, Kuhlen T (2007) Interactive Exploration of Large Data in Hybrid Visualization Environments. In: Proceedings of the IPT-EGVE Symposium, pp 69 – 76

[6] Schirski M, Kuhlen T, Hopp M, Adomeit P, Pischinger S, Bischof C (2005) Virtual

Tubelets — efficiently visualizing large amounts of particle trajectories. Computers & Graphics 29(1):17–27

[7] Wolter M, Bischof C, Kuhlen T (2007) Dynamic Regions of Interest for Interactive Flow Exploration. In: Proceedings of Eurographics Symposium on Parallel Graphics and Visualization, pp 61–68

[8] Hentschel B, Wolter M, Renze P, Schroeder W, Bischof C, Kuhlen T (2009) Hybrid Parallelization for Multi-View Visualization of Time-Dependent Simulation Data. In: Proceedings of the Eurographics Symposium on Parallel Graphics and Visualization, pp 79–86

[9] Cerfontaine P, Beer T, Kuhlen T, Bischof C (2008) Towards a flexible and distributed Platform for numerical Simulation. In: Proceedings of International Conference on Computational Science and Its Applications ICCSA 2008, LNCS, Springer Berlin/Heidelberg

[10] Schüppen A, Spaniol O, Thißen D, Assenmacher I, Haberstroh E, Kuhlen T (2008) Multimedia and VR Support for Direct Communication of Designers. In: Collaborative and Distributed Chemical Engineering. From Understanding to Substantial Design Process Support. pp 268–299, LNCS, Springer Berlin / Heidelberg

[11] Thain D, Tannenbaum T, Livny M (2005) Distributed Computing in Practice: The Condor Experience; Concurrency and Computation: Practice and Experience 17(2–4):323–356

[12] Schroeder W, Martin K, Lorensen B (2006) Visualization Toolkit: An Object-Oriented Approach to 3D Graphics. 4th edn. Kitware Inc.

[13] Assenmacher I, Hentschel B, Cheng N, Kuhlen T, Bischof C (2006) Interactive Data Annotation in Virtual Environments. In: Proceedings of Eurographics/ACM Symposium on Virtual Environments, pp 119–126

Authors

Since 2004, **Dipl.-Inform. Marc Wolter** has been a PhD student at the Virtual Reality Group, RWTH Aachen University, with Prof. Dr. Torsten Kuhlen. His research interests include scientific visualization, parallelization in interactive applications, psychological experiments in VR, and analysis of time-varying scientific data.

He has been actively publishing in VR- and visualization related conferences and scientific journals, and acted as reviewer for the IEEE Visualization and IEEE Virtual Reality conferences.

Thomas Beer received a B.Sc. Inf. and M.Sc. degree in computer science from Darmstadt University of Applied Sciences in 2003/2005. He is PhD student in the Virtual Reality group of the RWTH Aachen University since 2007 and works in the field of immersive visualization for mechanical engineering applications.

Dipl.-Inform. Philippe Cerfontaine, is PhD student in the Virtual Reality Group of the RWTH Aachen University since 2004. He specializes in the field of immersive visualization for the areas of virtual prototyping and mechanical engineering.

Bernd Hentschel got a Diploma (MSc.) in computer science from RWTH Aachen University in 2003. He is currently finishing his PhD with the Virtual Reality Group there. His research interests include scientific visualization in general, the use of information visualization techniques for simulation data analysis and feature-based visualization. He acted as reviewer for several international conferences, and the "IEEE Transactions on Visualization and Computer Graphics" journal.

Since 1997, **Prof. Dr. rer. nat. Torsten Kuhlen** has been the head of the Virtual Reality group at the Institute for Scientific Computing and the Center for Computing and Communication. Since 2000, he is CEO of the Virtual Reality Center Aachen (VRCA).

After his graduation in computer science in 1992, Torsten Kuhlen worked as a research assistant at the Chair of Computer Engineering of RWTH Aachen University with Prof. Kraiss. In 1998, he received his PhD with honors from the Faculty of Electrical Engineering and Information Technology.

He is a member of the Gesellschaft für Informatik (the German computer science organization); since 2006, he is the speaker of the section Virtual and Augmented Reality.

Torsten Kuhlen has been member of program committees of several Virtual Reality and visualization related conferences, as well as reviewer for journals and the German research foundation. He has written more than 150 publications for scientific journals, international and national conferences.

Corresponding author

Torsten Kuhlen(✉)
Virtual Reality Group, RWTH Aachen University, Seffenter Weg 23 52074 Aachen, Germany
E-mail:kuhlen@vr.rwth-aachen.de

Digital Olympic Museum and Sports Simulation

Zhigeng Pan[1], **Ruwei Yun**[2]
1. State Key Lab of CAD&CG, Zhejiang University
2. Edu-Game Center, Nanjing Normal University

Abstract

Digital Olympic Museum is one promising form to demonstrate the history, culture and highlights of Olympic Games. To provide functionalities of both sports and entertainment, several sports simulation systems are developed, which are Virtual Bowling, Virtual Pingpang, and Virtual Network Marathon. Virtual network marathon is implemented as one fitness-oriented virtual network game. Framework of prototype is illustrated. Three sports modes meet different users' requirements. Engine based scene and human edit enable users to deeply immerse into virtual environment and enable real-time rendering. Implementation of such system validates the feasibility to be used in digital Olympic museum.

Keywords

Olympic Museum, Virtual Reality, Real-Time Rendering, Sports Simulation

1 Introduction

In recent years, virtual reality (VR) is widely addressed in system design, toolkit development and some special applications etc., which refers to a technology that is capable of shifting a subject to a different environment without physically moving objects. VR is used in remote education, museum, sports, and in some cases it is combined with network for gymnastic fitness.

For a long time, physical museums have been one of the most important channels for transmission of knowledge and culture. People who visit physical museums browse directly the collections behind the glass boxes through text and picture, and are educating themselves. However, several limitations still exist for this form of physical demonstration, especially the constraints of time, space and interaction channels. With the aid of computer technologies and other advanced information technologies, digital museum has been developed as an efficient and promising alternative.

Conventionally, a digital museum is defined briefly as the digitalization of all the demonstrations in physical museum using extant digitizing technologies. These digitizing technologies include, but not limited to, panorama photography, videography, 3D scan, model reconstruction, audio recorder and network transmission. Thus, users are able to browse all the demonstrations anytime and anywhere. Collections of rare artifacts start to come into sight of general public. And more importantly, the mode of people browsing the digital museum is greatly changed from seeing what they are given, to seeing what they choose to see. Moreover, digital museum facilitates research, comprehensive communication and academic research all around the world.

Now, digital museum is widely used abroad, such as USA, Japan and Europe, where many museums are placed on Internet. In China, there are also many museums on Internet, whose content is simple, only a copy of text introduction, in addition to some stylized buttons, which links to the facade photos of collections[2–6].

When China was applying for holding the 2008 Olympic Game, we proposed a "Virtual Olympic Museum" in a creative way, which aroused the International Olympic Committee's great interest and attention, and it also benefits us a lot. After it was decided that China would hold 2008 Olympic Game, the central government planed to establish such a digital museum. The DOM system uses distributed virtual reality technology, which introduces the Olympic history and growing path in the background of diversified culture, which exerts great influence on all aspects of Olympic life. The visitor can roam the museum by controlling an avatar after login into the system. Meanwhile, he can communicate with surrounding users.

2 Digital Olympic Museum

The Olympic Games have a long history, in which one of the objectives is to lead to international friendship and understanding through competition in the sports stadium. This is commemorated in the physical Olympic museum in Lausanne, Switzerland, with a largely conventional website (www.olympic.museum.org). It is attractive to demonstrate the history, culture, highlights and sports gymnasium of Olympic Games in the DOM. Therefore, the two entities, virtual reality technologies and Olympic Games, motivate research into a virtual digital Olympic museum.

In order to implement the DOM system, we have researched and developed some techniques as follows: ① Storage and search of multimedia data: we use jpeg 2000 to code images[2,10,11], which supports transforming step by step. For video, we use MPEG-4, and media data content and semantic characteristic of the context to create index and finish searching. ② Modeling and rendering: we use level of detail (LOD) for rendering accelerating, and replacing geometrical modeling with image based modeling and rendering for real-time graphic generation[5]. ③ Simulation of sport and avatar: create simulation of basketball game and skeleton animation based on key frame.

2.1 System Structure

The DOM system uses techniques such as multimedia, network, distributed virtual reality, based on the model of Client/Server, as shown in Fig. 1. Therefore the differences between different hardwares of client can be shielded, reducing the complexity of system.

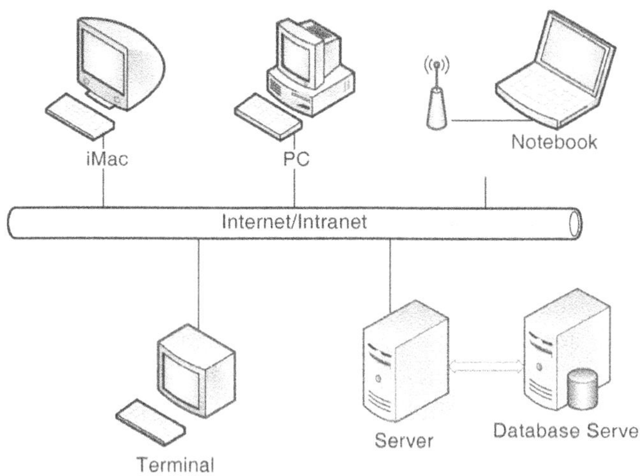

Fig. 1 The overall structure of DOM system

The server side is mainly for responding to requests from different clients, and providing services for them. It is composed of network service, database access interface, user management and service function (as shown in Fig. 2).

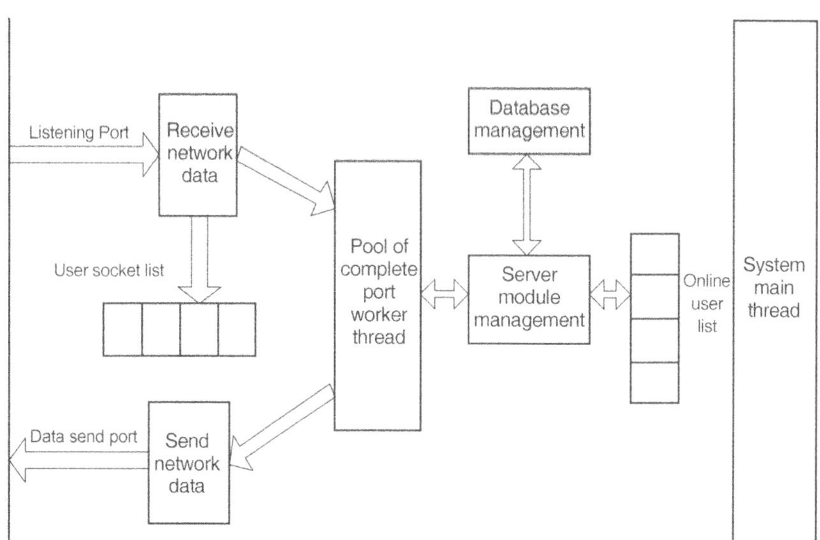

Fig. 2 The modules of server side

(1) Network service: The DOM system is based on TCP/IP protocol, and needs to save socket for every user. In order to get the effective server side, and increase the number of online users at the same time, we use complete port model (CPO), which is a service provided by the operation system kernel. The main ideal is organizing all the sockets into a pool, and generating a thread pool. Whenever there is a socket needing I/O request, the operation system picks up a thread to finish the job. This strategy increases the parallelism of DOM system greatly.

(2) Database access interface: In DOM system, we choose Mysql as background database, whose API is C, and we encapsulate it according to object orient thinking, and represent a database access session with three objects: connection, statement and result set. Meantime, we use database connection pool, which greatly saves the expense of creating then closing connection.

(3) User management: The server side needs to save all the online users, and divide them into groups according to different scenes, which makes searching easier, and the client rendering faster.

(4) Service modules: After analyzing the data from client, the server side needs to deal the data according to the request type, which includes: user login, user logout, scene switch, user synchronize, user communicate etc. Because of the diversity of the request, we use command design pattern14, which decreases the coupling between different modules of our system by separating the implementation and interface.

The client side is user oriented, and integrates with server side seamlessly, whose main modules include user interaction, scene management, network communication etc, as shown in Fig. 3.

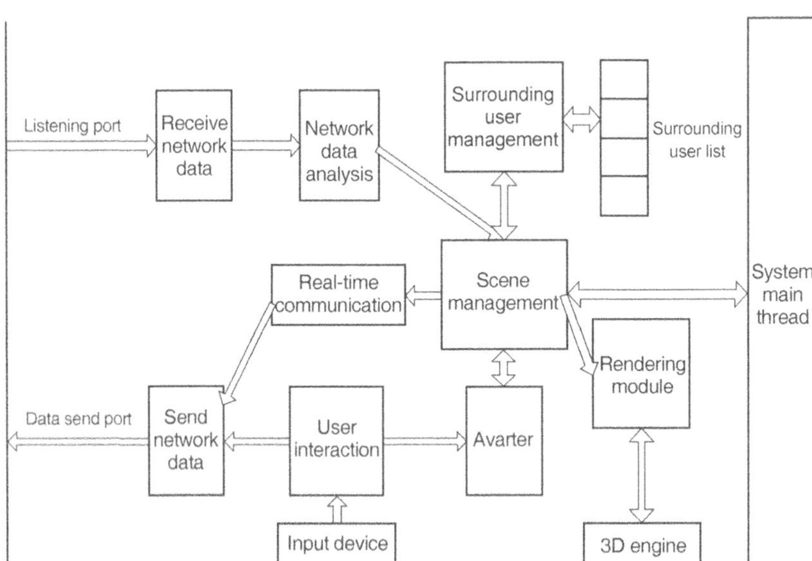

Fig. 3 The modules of client side

2.2 Storytelling in DOM

The DOM system tells users the knowledge about Olympic by presenting objects from Olympic museum in 3D form, in addition to some special elements of China, as shown in Table 1.

Table 1 The knowledge included in the digital museum

Content	Description
Olympic culture	Introductions about flag and symbol, torch and flame, anthem, molto, medals, etc.
Olympic history	Displays and introduces Olympic antiques, the holding places and time of all the previous Olympic games
Olympic objects	Displays torch, medal, poster, mascot of each Olympic Game in 3D form
Sports'culture of China	The growing path of gymnastic movement in China and the cultures in each stage

According to the knowledge included in the digital museum, we separate it into four parts, as shown in Fig. 4.

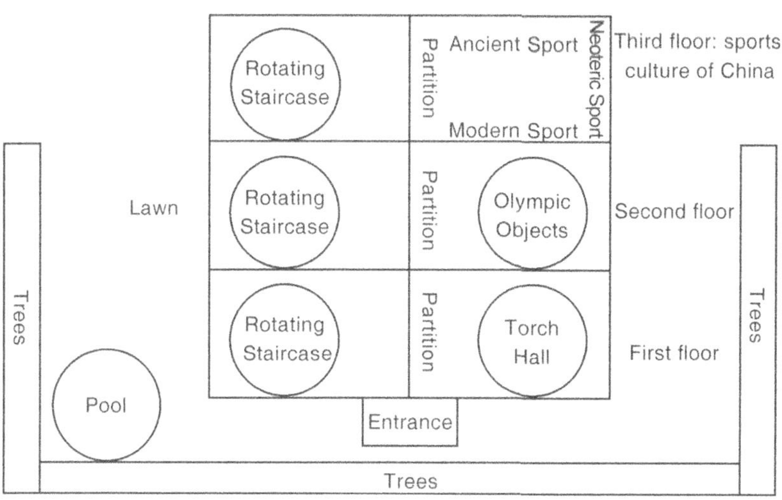

Fig. 4 The architecture of Olympic knowledge

The surrounding displays the landscape around the museum, including the main building, lawn, tree, shrub, street lamp, water pool and steps. Users can enter the museum through the entrance of the main building.

The first floor is divided into two parts. The first part is rotating staircase, by which user can walk to the second floor, and there are several pillars distributed around the staircase, on which words about Olympic knowledge are carved. Users can see the torch hall after crossing the partition, which presents the Olympic torches and corresponding introduction. Meanwhile, we put some cultural relic near the wall, and posters of all summer Olympic on the wall to increase the vitality.

The second floor includes two parts. The first is the staircase connecting the first floor with the third floor. The second floor is mainly used for presenting Chinese sports culture, including three parts: ancient, neoteric and modern. This part embodys the development path and details of Chinese sports, and the far-reaching significance.

The third floor can also be divided into two parts. The first part is the entrance back to second floor. We put the photographs of Olympic Committee presidents on the wall, along with their personal introduction. Users can arrive at the exhibition hall after crossing the partition, which uses windows as main structure. We put Olympic trophies, medals, mascots, Beijing Fuwas and so on inside the show windows. Near the wall, we place a number of antique artifacts related to Olympic Games.

2.3 Modeling and Rendering of DOM

The DOM system digitalizes the Olympic museum in the real world, so we need to build a large number of 3D models according to the real one. The modeling includes two aspects: 3D object modeling and physical factors modeling.

We use 3DS Max to finish 3D object modeling15. According to the scene design, we create models for each floor respectively. During the modeling, we use the framework provided by DirectX Shader to achieve some advanced effects, such as transparent textures and metal materials. After that, we export every model into XML file with the Lucid3D16 exporter. Each XML file includes vertex coordinate, normal, material, texture coordinate, as well as effect files.

The physical factors refer to collision detection, light in 3D scene, so physical modeling includes setting lights and placing collisions. With the help of scene builder17, we can merge the scenes, place directional lights, point lights and spot lights easily, meanwhile, we can also set collisions for each model in the scene, so the avatars never get intersection.

In DOM system, we need to ensure the accuracy of the models, and avoid being felt rough and deformed on one hand. On the other hand, we need real-time rendering. As both of them are contradicted with each other, we need to search for balance between them. With several optimization technologies, we gain a relatively good result.

(1) Scene segmentation. By dividing the big scene into small scenes, we can greatly reduce the rendering resources needing to be loaded each time. Whenever the avatar enters the floor, we reload the scene resource.

(2) Model optimization. We remove the invisible faces, such as internal structure of a model and models outside the perspective. For complex sharps, complex or coincident surface, we simulate with contour and simple outline respectively.

(3) Texture optimization. We use image rendering from high-precision model as texture for low-precision model. Meantime, we change the size of texture into exponent of 2, and use the dds format texture which supports mip-map.

The DOM system tells users the knowledge about Olympic by presenting objects.

2.4 Experimental Results

The system is implemented based on Lucid engine[1,12] from the Hong Kong Polytechnic University. In Fig. 5(a), we can see the main building, the lawn, the steps to the museum and roads. Figure 5(b) is the actual landscape. Figure 6 (a) is the torch hall, and (b) is the actual one.

(a) (b)

Fig. 5 The surrounding landscape of DOM (a) and the actual one (b)

In Fig. 6(a), we can see the torches hall, which displays torch of each Olympic Game and their responding introduction. Figure 6(b) is the actual torch hall.

(a) (b)

Fig. 6 The torch hall of DOM (a) and the actual one (b)

In Fig. 7, we can see the displays and introduction on the ancient sport culture of China, from which we can learn more about ancient China.

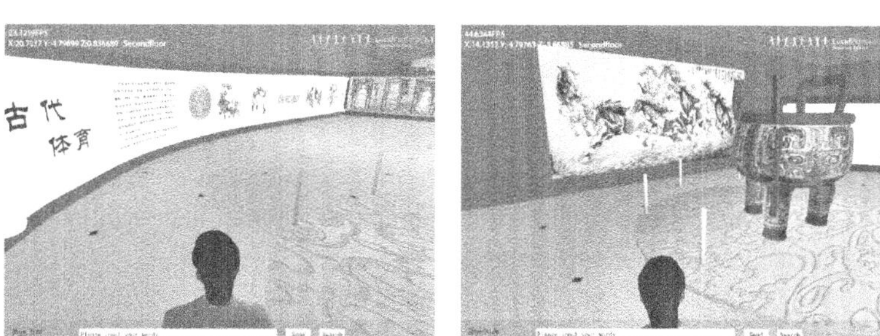

Fig. 7 The ancient sport of China

In Fig. 8, we display some pictures and introduction on neoteric gymnasium (a) and modern champions in the Olympic Games (b).

(a) (b)

Fig. 8 The neoteric sport of China (a) and modern champions in the Olympic Games (b)

3 Virtual Reality Based Sports Simulation

Sports simulation is one kind of systems engineering science[6,9]. Through technologies of computer simulation, it reproduces teaching experience of teachers, training intention of coaches, administrators'organization scheme and athletes' training process. Sports simulation helps to interpret, analyze, predict, organize and evaluate one sports system. As the advanced human-computer-interaction technique, virtual reality provides user with more sense of immerse, interactivity and imagination. Correspondingly, virtual reality based sports simulation satisfies both athlete and coach, and other participants as well, which is widely used in the field of sports.

3.1 Main Functionalities and Key Technologies

Virtual reality based sports simulation systems generally have the following functionalities:

- Constructing of virtual training scene;
- Capturing motion data;
- Collecting physiology, biochemistry and psychology data;
- Reproducing action;
- Illustrating and analyzing training effects.

Correspondingly, key technologies of sports simulation include physics and physiology based modeling, virtual human animation, motion data capture, real-time rendering and nature human-computer interaction.

3.2 Typical Prototypes

Now, DOM is extended from traditional Olympic Games into sports types and functions.

Some mass sports are also integrated into VDOM. Several virtual reality based sports simulation systems have been successfully implemented, such as Easybowling and virtual network marathon (VNM). They both integrate body exercise into game playing.

Easybowling. Players throw a real bowling ball on a 2-meter-long track, and then the EasyBowling system uses a PC camera to detect the speed and direction of bowl. After the motion parameters are computed, the movement of the bowling ball and its collision with pins are simulated in real-time and the result is displayed on a large display screen. Since it only requires a small space, this real game mode machine can also be a family game machine. Figure 9 shows the prototype of Easybowling[7].

Fig. 9 Prototype of Easybowling (left) and collision simulation (right)

Virtual network marathon. Users do physical exercise and competition in a distributed virtual environment[8]. When users run on the elliptical machine, their body motions are captured by sensors. They can observe their actions and current environment real-time rendered through output screen, and control their actions by multimodal interactions or interactive with other users by network. There are three game modes, training, exercising and competition. Users are also able to choose the appropriate distance to run. Figure 10 shows the exercise modes of VNM (walkthrough in museum).

Fig. 10 Exercise modes of virtual network marathon

VR-based Basketball teaching. As is known, modern basketball continuously improves during the mutual promotion and restriction between basketball technique and basketball rule. Teachers and coaches who are engaged in popularization and promotion of basketball sport cannot always demonstrate normative technical movement. Even for leading players,they might face the same problem with their age increasing and degradation of technical and physical quality. In order to achieve correct technical movement, the feedback process is accomplished by students' repeated trainings and teachers' guidance. Sometimes, due to complex tactic content, teachers cannot make out the movement directions and opportunities of tactics in lots of cross line graphs. Many teachers analyze tactic line graph through experiences. Basketball technique and tactics simulation is based on VR means applying VR technology to simulate basketball sport. As a brand-new technology, it provides a new idea and teaching platform for modern basketball teaching and training. This simulation does not only enable us to view and display scientific, reasonable, normative technical movement by breaking time-space limitation at any angle, but also displays tactical coordination line dynamically. All of these can avoid the defects due to traditional teaching and training demonstration. (Fig. 11 is an example).

Fig. 11 Training of Basketball (animation clips)

4 Conclusion and Future Work

The DOM system provides end-users with the means of navigating the Olympic museum, which has a number of restrictions on the form of display. In addition, it also presents the knowledge source on Olympic games and sports culture of China in a human-oriented and effective way. And we also provide end-users with some freedom to navigate the museum.

However, there is still a long way to go before implementing a complete DOM system, as the following:

(1) Support multiple avatars. As the number of online users grows and the rendering rate becomes slower, we need some ways to solve this problem.

(2) Increase immersion. There is no sound in current system. In order to add sound effect in the scene and increase the feeling of immersion, we plan to add sound during the rendering of the scene.
(3) Integrate information retrieval module. We will add text and image search module into current system, which was developed by Nanjing University.

Acknowledgments

This project is co-supported by Key NSFC (grant no: 60533080) and 863 project in China (grant no: 2006AA01Z303). Many people contribute to these projects; they are Mr. Zhongfu Li, Mr. Zi Li and Mr. Yuhu Ma for modeling of the museum; Dr. Gaoqi He, Mr. Yongqing Li, Mr. Mingliang Xu, Mr. Pei Lv for implementing the NVM system; Mr. Haibo Yuan, Mr. Jianfeng Liu, Mr. Weiming Pan for implementing the DOM. These systems are implemented on the game engine called Lucid, which is provided by Prof. Gino Yu in The Hongkong Polytechnic University. The retrieval part is implemented by Prof. Gangshan Wu and his students in Nanjing University.

References

[1] Cheung M, Tsui B (2005) Lucid3D artist guide. The Hong Kong Polytechnic University, Hong Kong
[2] Cui C, Shi J (2004). Analysis of feature extraction in 3D model retrieval. Journal of Computer-Aided Design & Computer Graphics 16(7): 882–889
[3] Gray G (1999). Discussion on future directions of global digital museum: Museum and GUI. In: S Sugita, Hong, J K, J Reeve et al(eds) Senri Ethnological Report, Global Digital Museum for Museum Education on the Internet
[4] Gray G (2001). Co-construction of digital museums. Spectra 28(1):12–14
[5] Hoppe H, DeRose T et al (1992). Surface reconstruction from unorganized points. In: ACM SIGGRAPH 1992 Conference Proceedings, pp71–78
[6] Ji Q, Pan Z et al(2004). Virtual arrangement and rehearsal of group calisthenics. Journal of Computer-Aided Design & Computer Graphics.16(9):1185–1190
[7] Pan Z G, Xu W W (2003). Easybowling: a small bowling machine based on virtual simulation.Computers & Graphics 27(2):231–238
[8] Usaka T (1996). A large scaled multi-user virtual environment system for digital museums.In: Division of Science, Graduate School of University of Tokyo
[9] Xu W, Pan Z et al (2003) Footprint sampling based motion editing. International Journal of Image and Graphics 3(2):311–324
[10] Zhang Z, Pan Z et al (2005) Multiscale generic fourier descriptor for gray image retrieval. Journal of Image and Graphics 10(5):611–615
[11] Zhang Z, Yang B (2006) Visual projection similarity for matching similarity

among 3D models. Journal of Computer-Aided Design & Computer Graphics 18(7):1049–1053

[12] Yip M K, Cheung M, Cheung P C et al (2005) Lucid3D programmer guide. The Hong Kong Polytechnic University, Hong Kong

Author

Prof. Dr.Zhigeng Pan received his Bachelor Degree and Master Degree from the Computer Science Department in 1987 and 1990 from Nanjing University respectively and Ph.D Degree in 1993 from Zhejiang University. Since 1996, he has been working at the State Key Lab of CAD&CG as a full-time professor. Dr. Pan is a member of SIGGRAPH, Eurographics, IEEE, a senior member of the China Image and Graphics Association. He is on the director board of the International Society of VSMM (Virtual System and Multimedia), a member of IFIP Technical Committee on Entertainment Computing (acting as representative from China). Currently, he is the Editor-in-Chief of The International Journal of Virtual Reality. He is on the editorial board of International Journal of Image and Graphics, International Journal of CAD/CAM, Journal of Image and Graphics, Journal of CAD/CG, He is the program co-chair of EGMM'2 004 (Eurographics workshop on Multimedia), Edutainment'2005 and VEonPC'2005, and is the program co-chair of Edutainment'2006, conference co-chair of ICAT'2006, Cyberworlds2008.

Corresponding author

Zhigeng Pan(✉)
State Key Lab of CAD&CG, Zhejiang University, Hangzhou, 310027, China
E-mail: zgpan@cad.zju.edu.cn

Research on Key Technologies of Full Mission Navigation Simulation System

Yong Yin, Yicheng Jin, Hongxiang Ren, Xiufeng Zhang, Xiuwen Liu and Junsheng Ren

Key Lab. of Marine Simulation & Control, Dalian Maritime University

Abstract

As an important application field of virtual reality and a typical man-in-loop simulation system, how to evaluate the performance of Full Mission Navigation Simulation System is very important. The paper presents three realism specifications to evaluate a Full Mission Navigation Simulation System: behavioural realism, environmental realism and physical realism. The three realism specifications are related with the key technologies used in the Full Mission Navigation Simulation System, which are system architecture, ship hydrodynamic mathematic model, visual system and simulated equipments. Behavioural realism is determined by ship hydrodynamic mathematic model, environmental realism is mainly determined by visual system and physical realism is determined by the simulated bridge and equipments. The paper introduces functions and requirements from three kinds of evaluation realism and key technologies in details.

Keywords

Full Mission Navigation Simulation System, Navigation Simulator, Virtual Reality, Computer Graphics, Application of Simulator

1 Introduction

International Convention on Standards of Training, Certification and Watch keeping for Seafarers 1978 (amended in 1995) of International Maritime Organization (IMO) (STCW 78/95)[1] has been brought into effect on 1st, Feb 1997. More contents about using simulator in maritime education, training and certification are added in the convention amended in 1995. As one of the council countries of IMO, China has responsibility to perform various international convention of IMO; therefore, navigation colleges, schools and training center of China must buy and install a lot of navigation simulators. We began to research the simulator from the end of 1980s because of the expensive price of imported simulator. The aims of the research are to prevent the market monopolization of simulator, increase the technical level of our country in marine simulator research, development and application in ocean engineering, and providing a successful application example of virtual reality. Through about two decade's efforts, we have developed Radar/ARPA (Automatic Radar Plotting Aid) simulator, ship handling simulator and full mission navigation simulation system successfully. The specification of the simulator is based on the Standard for Certification No.2.14 of Maritime Simulator Systems[2] established by DNV, a famous ship classification society from Norway. In the standard, three essential realism specifications are used to evaluate a Full Mission Navigation Simulation System: behavioural realism, environmental realism and physical realism. Behavioural realism is determined by ship hydrodynamic mathematic model; environmental realism is mainly determined by visual system; physical realism is mainly determined by the simulated bridge and equipments. DNV certification standard puts forward a very high requirement of simulator objectives and three realisms. The paper will introduce the key technologies used in the Full Mission Navigation Simulation System in details: system architecture, ship hydrodynamic mathematic model, visual system and bridge equipments simulated.

2 Key Technologies

The specification of full mission navigation simulation system is based on the Standard for Certification No.2.14 of DNV. Based on rapid development of computer technology, simulation technology, computer graphics technology, virtual reality and the right estimation of developing trend of these technologies [3], we brought forward a reasonable architecture and feasible technology scheme.

2.1 Architecture of Full Mission Navigation Simulation System

The architecture of Full Mission Navigation Simulation System adopts the design idea of Distributed Interactive Simulation (DIS) and High Level Architecture (HLA) [4, 5]. The system consists of instructor operation station(IOS), one primary ownship with 270 degrees visual system, several secondary ownships with 180 degrees visual system and CCTV system for monitoring the operation of trainee. Figure 1 is about the architecture of the system. High performance PCs are used in the IOS and each ownship. Computers are linked by network. In primary ownship, there is a mimic ship wheelhouse, control devices and visual system with 270 degree field of view. Cylindrical screen seamless tiling projection image using software edge-blending technique with wide field of view is used in the visual system. Figure 2 is about the system architecture of primary ownship. The technologies and configuration used in secondary ownships are the same as the above ones used in primary ownship. If considering the cost of the whole system, the configuration of secondary ownships can be decreased. Because there is multi-ownship with different level configuration in the system, the system can fulfil the requirements of anti-collision of multi-ship and can increase the number of trainee; thus, it can adapt to the situation in our country; namely, many seafarers waiting for being trained.

Full Mission Navigation Simulation System adopts some advanced technologies, such as computer generating image technology, virtual reality and curve screen seamless tiling projection using software edge-blending technique with wide field of view[6–17]. It can provide operators a realistic virtual enviroment with digital harbour and machinery in harbour. The ship hydrodynamic mathematic model using separating method[17–24] can meet the requirement of training and competence certification, and can be used in the demonstration of ocean engineering[25] and the analysis of maritime disaster.

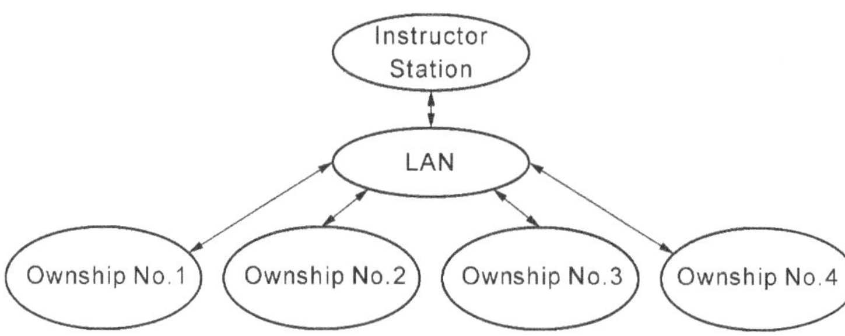

Fig. 1 System architecture

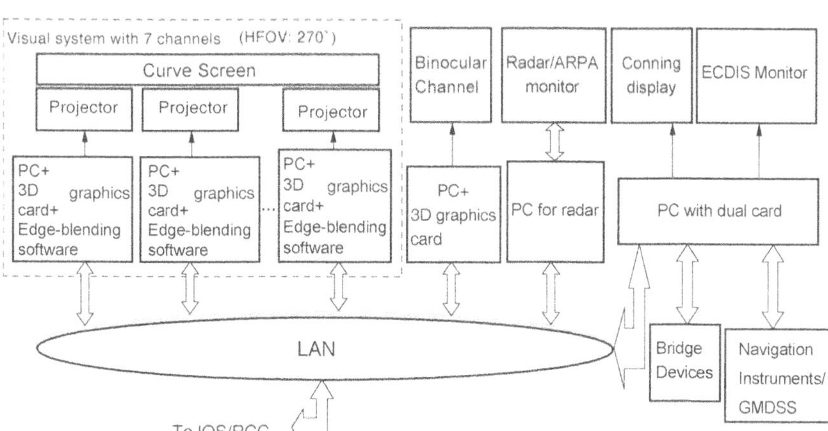

Fig. 2 Configuration of primary ownship

Figure 3 is about the mimic ship bridge wheelhouse. Some simulation equipments similar as in real ship are configured in the bridge, such as Electronic Chart Display Information System (ECDIS), RADAR/ARPA, throttle, rudder, communication equipments and displaying instruments. The bridge is surrounded by a cylindrical screen with 270 degree field of view. Operator can see the outside image generated by computers and projected on the screen through the window. Operators within the virtual environment can look a realistic scenery with day/night changing and different visibility. They also can feel the movement of ship with 6-DOF and drive the virtual ship in the virtual harbor by operating the equipments installed in the ship bridge simulated, such as throttle, rudder, thruster, anchor and communication equipments.

Fig. 3 Mimic ship bridge (inside)

The bridge construction of primary ownship is shown in Fig. 4. The right-bottom is the bridge wheelhouse. Operator will be in the wheelhouse while training. They can see the outside ocean scenery generated by visual system through the window of bridge. The radius of cylindrical screen is 8 meters. The horizontal field of view is 270 degrees.

Fig. 4 The construction of ship bridge and visual system

Seven projectors are installed over the bridge. The images generated by computer must be processed by geometry correction and edge blending module before they are projected on the screen for providing operator a seamless image with wide field of view.

Figure 5 is about one of the pictures of secondary ownship for simulating the bridge of a patrol ship. Three-channel visual system is used in the ownship for displaying ocean scenery with 120 degree FOV. The radius of cylindrical screen is 5 meters. The equipment in the mimic bridge is the same as a real bridge. Figure 6 shows the layout of instructor operation station(IOS). IOS can control the running of whole simulation system, mainly consisting of creating and editing exercise, choosing ship type and loading condition, setting environment condition such as wind, current, tide and visibility, setting failure modes, monitoring each ownship and replaying after training etc.

Fig. 5 Secondary ownship

Fig. 6 Instructor operation station

2.2 Ship Hydrodynamic Mathematic Model

The behavioural realism of simulation system is determined by the 6-DOF ship hydrodynamic mathematic model when ship navigates in ocean including horizontal motion, vertical motion models and responses when operators manoeuvers ship by operating some control devices such as throttle, rudder, anchor, mooring line, tug boat [18]. From the viewing degree of ship principle, horizontal motion model of ship belongs to the category of ship manoeuvering prediction, where as vertical motion model of ship belongs to the category of wave-resistance prediction. The methods of ship manoeuvering prediction include database method, self-navigation model test method and numerical simulation of ship motion mathematic model[19–24] etc. Here, ship manoeuvering prediction using ship hydrodynamic model is implemented by combining computer simulation technology and the hydrodynamic derivatives through constrained ship model test. The method has been recognized in "Ship Manoeuvering Standard" (IMO convention MSC.137(76)). For full mission navigation simulation system, the ship manoeuvering mathematic model can be used to implement different training items, such as berthing and deberthing, navigation in narrow channel and bad weather condition, anchorage operation, besides manoeuvering prediction; therefore, all the requirements should be taken into consideration while applying the model in simulation system.

By now, the manoeuvring mathematical model of ship can be classified into two types:

one is hydrodynamic model (including whole model and separating model) and the other is response model (including two-order model and one-order model). The separating model has some advantages such as the clear physical meaning of parameters in the model, the parameters being calculated conveniently, model precision is higher. In addition, it can simulate different ship types and meet the requirements of ship berthing and de-berthing. Therefore, the method of separating model is adopted in the simulation system.

The 6-DOF mathematical model is adopted 4+2 structure in our research. One part is 4-DOF model to simulate the couple of ship's motion in plane and ship's rolling motion; the other is 2-DOF model to simulate the couple of heaving and pitching. This method can decrease the complex degree and enhance the robust of the model. The following 7 researches have been done:

(1) Ship hydrodynamic model with 6-DOF[18]. The 6-DOF motion of ship can be implemented by 6-DOF motion platform and visual system. To increase the ratio of performance and price, we adopt visual system to provide the 6-DOF motion feeling.

(2) Ship hydrodynamic model in open water, including the influence of wind, current, tide and ship loading condition.

(3) Ship hydrodynamic model in restricted water, including shallow water effect, bank effect and ship-ship interaction. The low-speed model with four quadrants can meet the requirement of berthing and de-berthing training [19–24].

(4) Abundant ship models have been created including more than 150 models with different types such as oil tanker, container ships, bulk ship and passenger ship.

(5) Hydrodynamic models of ship equipped with different propeller such as CPP (Controllable Pitch Propeller), VSP (Voith Schneider Propeller) and Water jet ship [19–24].

(6) Study the developing platform of ship hydrodynamic model. The platform is based on a lot of sea trial data of real ship. By using regression analysis method, different ship model can be created with a higher precision by revising the ship's parameters conveniently [19–24] even if the ship's parameters and sea trial data are not enough.

(7) Design some autopilot algorithms and their debugging platforms: course-keeping, track-keeping, etc[26].

2.3 Visual System

The operating environment in simulator is a virtual environment which lets operator feel immersion. It mainly includes visual system, sound system and motion sense. In recent years, the progress of virtual environments development is the most rapid part of navigation simulator. STCW 78/95 Convention gives some recommended guidance to some non-compulsory training or competence certification items such as bridge resource management training. In STCW, simulator for navigation and on-duty training should "provide a realistic day or night scenery, including various visibility; or whenever observing in ship bridge, it can provide trainee a minimum horizontal FOV in visible area which meets the task and aims of navigation and on-duty." The realistic scenery here mainly includes the following contents and requirements:

(1) Sky doom: sunshine, cloudy and foggy[27].

(2) Weather condition: sunshine, raining and snowing, lightning[28–29] .

(3) Sea: real 3D movement sea surface with texture, wave height and period is correlative with wind force and wind direction. The parameters of wave are calculated according to wave spectrum. Ship's 3D model moves with wave[30–35].
(4) Terrain: real 3D terrain models based on real terrain of harbor and real textures, more realistic[36–37] .
(5) Wharf: including building, cranes, bollards, vehicles, etc.
(6) Navigation Aids and lights: visible distance and flashing periods of navigation aids are compliant with IMO rules.
(7) Typical buildings in harbor: all building is mapped with real day and night textures[33, 38].
(8) Bow wakes and ship's trails: all of them are correlative with ship'speed and course, move with wave and fade with time[30] .
(9) Navigation lights of target ship: IOS can control the status of all lights of target ships. At night, the visible distance and scope of each light are compliant strictly with IMO rules.
(10) Simulation time: changes continuously from dawn, day and dusk to night.
(11) View point switch: the view point can change to different position such as bow, stern, deck or in the sky for displaying birds'-eye view[39, 43].
(12) Different special effect display: such as oil diffuse, fire, explosion, etc.

Obviously, the key technologies of visual system mainly focus on the realism and real-time ability of scenery. Because the graphics workstation is more expensive than PC with graphic card, in order to keep the simulator competitive in the market, the full mission navigation simulation system adopts CGI technology, and PC-cluster architecture to generate image. As most commercial modeling software and scene management software can not implement all of the functions or the effect is not satisfied, we have developed some software to generate visual database such as:

(1) Coordinate conversion software of Gauss projection and Mercator's projection.
(2) Real-time wave simulation software based on wave spectrum for navigation simulator[31, 34, 35].
(3) Simulation software of ship's wakes and trails based on particle system[30].
(4) Rain and snow simulation software based on particle system[29].
(5) Simulation algorithm of navigation aids such as buoys, lighthouse and ship's navigation lights.
(6) Binocular simulation software[39, 43].
(7) Terrain Modeling software based on contour lines of chart[36, 37].
(8) Software synchronization algorithm of multi-channel system based onnetwork [15].
(9) Geometry correction and edge blending software of multi-channel projection system[16, 40, 41].
(10) Real-time shadow rendering software a scenery of large scale[32, 42, 45].

By now, we have created more than 30 harbor visual databases including Sydney and Singapore harbor by using the software above. Geometry correction and edge blending software of multi-channel projection system is developed by ourselves[16]. The software maps the image generated in the graphic pipeline to a projection surface as dynamic texture in real-time. The work is finished within the graphic card directly before the image is transferred to projector. Using this technology, we can use normal projectors in the multi-projection system with cylindrical screen. The effect of geometry correction and edge blending is much better than that of hardware correction because it almost has no failure with lower cost.

Because the visual system of navigation simulator adopts mostly single view point, the depth cue is not good. Although increasing the radius of screen can improve the depth effect, the requirement of big room and screen will require more investment. We have developed a visual system using stereo system to recover the depth information through the parallax scene between two eyes, in order to greatly enhance the immersion effect of virtual environment[40, 41].

Fig. 7 Sample frame from simulator

2.4 Mimic Ship Bridge and Equipments

Physical realism is determined by the simulated bridge and equipments. Navigation simulator is a typical man-in-loop simulation system. Operators can participate in the simulation system, and control the system according to various information provided by the system. According to STCW 78/95 convention, the physical equipments in simulator "can simulate the operation performance, meet the training aims and simulate the limitation and possible error". With the rapid development of virtual reality, the simulation of equipments can be implemented by two ways: one is physical equipments such as real throttle, rudder and console; and the other is virtual equipments, where operatormust put on a head-mounted device, body tracking system and data glove. Because ship navigation is also based on Bridge Team Work now, there are several persons in ship bridge simultaneously when ship is navigating. They must cooperate with each other, but by now VR technology can not merge the images of colleagues who are working together in one virtual image; therefore,we choose semi-physical simulation to implement the navigation simulation, such as mimic ship bridge, real control devices (throttle, rudder, compass and meters, etc.), wide-FOV visual system using projector and sound system.

In the system, we have built a mimic ship bridge wheelhouse which is the same as a real ship's and installed similar equipments or simulative system in the bridge, mainly including:

(1) Remotely control devices of main engine, such as throttle, CPP control unit and thruster control panel.
(2) Remotely control of auxiliary engine, such as generator control.
(3) Rudder control system, including manually operating rudder and autopilot.
(4) Compass indicator for rudder operation and azimuth compass.
(5) Radar and ARPA simulator[6, 12, 44].
(6) ECDIS.
(7) Main navigation devices, such as GPS, AIS, echo sounder and speed log [46, 47].
(8) GMDSS (Global Maritime Distress Safety System) equipments [48, 49].
(9) Intra-communication in ship.
(10) Tug boat operation.
(11) Anchor operation.
(12) Mooring line operation.
(13) Navigation light operation.
(14) Deck light operation.
(15) Sound control according to rules of anti-collision.
(16) Meter indicators: wind force and direction, rudder angle, rate of turning, depth indicator, speed meter and engine RPM (Revolutions Per Minute), etc.

3 Conclusion

Navigation simulator is a typical man-in-loop simulation system and also an important application field of virtual reality. Three realism specifications are used to evaluate a Full Mission Navigation Simulation System: behavioural realism, environmental realism and physical realism. Behavioural realism is determined by ship hydrodynamic mathematic model, environmental realism is mainly determined by visual system and physical realism is determined by the simulated bridge and devices.The paper introduces the key technologies used in the Full Mission Navigation Simulation System: system architecture, ship hydrodynamic mathematic model, visual system and equipments simulated. The system builds 6-DOF hydrodynamic mathematic model for each simulating object-"Ownship". The model will calculate ship's movement parameters in real-time according to real ship's particular data, navigational environment and control command when simulator has run. The motion parameters will be transferred to visual system and console for displaying such as meters and conning display. After visual system has received the parameter, it can calculate visual parameters and generate 3-D virtual environment. PC-cluster and multi-projector are used in the visual system for generating and displaying the virtual environment. Seamless tiling technology using edge blending and geometry correction is adopted in the cylindrical screen projection system for providing a strong immersion feeling. When operators locate in the virtual environment, they can feel the movement of ship with 6–DOF and drive the ship in the virtual harbor by operating the equipments installed in the ship bridge wheelhouse simulated, such as throttle, rudder, thruster, anchor and communication equipments, etc.

References

[1] International Maritime Organization (IMO) (1995) International convention on standards of training, certification and watch keeping for seafarers 1978 (amended in 1995)

[2] DNV (2000) Standard for certification No.2.14 maritime simulator systems

[3] Shi J Y(2002) Virtual reality basic and algorithm. Press of Science, Beijing

[4] Yin Y, Jin Y C (2000) Navigation Simulator and Distributed Interactive Simulation Technology (In Chinese). Journal of Computer Simulation 17(6): 66–68

[5] Yin Y, Jin Y C, Li Z H (2000) Network communication in distributed navigation simulator (In Chinese). Journal of System Simulation 12(6): 621–624

[6] Yin Y, Jin Y C (1996) Study on image generation in radar simulator using page update method (In Chinese). Journal of Dalian Maritime University 22(4):31–36

[7] Yin Y, Jia C Y (1996) Study on an algorithm for acquiring three dimensional information used in ship handling simulator (In Chinese). Journal of Dalian Maritime University 22(2): 14–18

[8] Yin Y, Jin Y C (1997) Research strategy of multi-function navigation simulator (In Chinese). In: Proceedings of Annual Conference of Chinese System Simulation Association

[9] Jin Y C, Yin Y (1997) Study on visual system developed using openGL (In Chinese). In: Proceedings of Annual Conference of Chinese System Simulation Association

[10] Jin Y C, Yin Y (1998) Study on night scene and foggy scene in navigation simulator (In Chinese). Journal of Dalian Maritime University 24(2): 26–31

[11] Yin Y, Jin Yicheng (1998) Economic multi-ownship and multi-function navigation simuator (In Chinese). Journal of Dalian Maritime University 24(2): 16–21

[12] Yin Y, Jin Y C (1998) Radar image generation of navigation radar/ARPA simulator using PC (In Chinese). Journal of Shanghai Maritime University 5:107–111

[13] Jin Y C, Yin Y (1999) Study on visual system of navigation simulator (In Chinese). Journal of Dalian Maritime University 25(2): 22–27

[14] Jin Y C, Yin Y (2002) Development of STCW convention and navigation simulator (In Chinese). Journal of Dalian Maritime University 28(3):51–55

[15] Yin Y, Jin Y C, Li Z H (2002) Research on key technologies of multi-channel visual system. In: The 5th International Conference on System Simulation and Scientific Computing (Shanghai) (ICSC'2002), pp 436–439

[16] Liu X W, Xie C, Jin Y C et al (2006) Construct low-cost multi-projector tiled display system for marine simulator. In: 16th International Conference on Artificial Reality and Telexistence, ICAT 2006, Hangzhou, China, November 29 –December 1, 2006. Workshops Proceedings. IEEE Computer Society 2006, pp 688–693

[17] Yin Y, Ren H X, Jin Y C et al (2005) Research on simulation system based on virtual reality technology for handling underwater vehicle in deep sea (In Chinese). Journal of Harbin Engineering University 26(4):435–440

[18] Zhang X F, Yin Y, Jin Y C (2007) Research on ship motion mathematical model with six degrees of freedom in regular wave (In Chinese). Journal of Communication Transportation Engineering 7(3):40–43

[19] Sun X F, Yin Y, Zhang X F (2007) Maneuvering mathematical model for ships equipped with controllable pitch propeller (In Chinese). Journal of Dalian Maritime University 33(4):73–76

[20] Sun X F, Yin Y, Zhang X F (2007) Mathematical model of mooring lines in ship handling simulator (In Chinese). Journal of Navigation 72(3): 1–4

[21] Ren J S, Yin Y, Zhang X F (2006) Study on maneuvering model of kamewa-type waterjet ship for shiphandling simulator. In: Proceedings of International Conference on Marine Simulation and Ship Manoeuvrability (Marsim2006), Terschelling, Netherlands, June 2006, P: M–20–1~M–20–5

[22] Sun X F, Yin Y, Zhang X F (2006) Study on maneuvring mathematical model for trawler. In: Proceedings of International Conference on Marine Simulation and Ship Manoeuvrability (Marsim2006), Terschelling, Netherlands, June 2006,P:M–30–1~M–30–5

[23] Zhang X F, Jin Y C, Yin Y et al (2005) Propeller model and use in the tug moving simulation. In: Proceedings of Asia Simulation Conference/the 6th International Conference on System Simulation and Scientific Computing (ICSC'2005), Beijing, 2005. International Academic Publishers World Publishing Corporation 10:214–218

[24] Sun X F, Yin Y, Zhang X F (2007) Research on Dynamic Model of Trawl Warp in Fishing Simulator. Journal of System Simulation (CCVRV2007) 19(2):14–16

[25] Zhang X F, Jin Y C, Yin Y et al (2004) Ship simulation using virtual reality technique. In: Proceedings VRCAI 2004–ACM SIGGRAPH International Conference on Virtual Reality Continuum and Its Applications in Industry, Proceedings VRCAI 2004–ACM SIGGRAPH International Conference on Virtual Reality Continuum and Its Applications in Industry, 2004, pp 282–285

[26] Yin Y, Zhang X F, Zhang X K(2007) Research on simulation platform for testing algorithms of autopilot and automatic collision avoidance (In Chinese). Journal of System Simulation 19(1):56–58

[27] Shen H L, Yin Y(2008) Real-time dynamic simulation of 3D cloud for marine search and rescue simulator. In: ACM SIGGRAPH International Conference on

Virtual-Reality Continuum and Its Applications in Industry (VRCAI 2008), Dec. 2008

[28] Yin Y, Jiang J (2007) Modeling and simulation of celestial scene in shiphandling simulator (In Chinese). Journal of Dalian Maritime University 33(4): 42–46

[29] Yin Y, Jin Y C, Ren H X et al (2002) Real-time simulation of natural phenomena (In Chinese). Journal of System Simulation 14(9): 1217–1219

[30] Yin Y, Ren H X, Zhang X F et al (2002) Wave simulation technology applied to virtual environment of navigation simulation (In Chinese). Journal of System Simulation 14(3): 311–315

[31] Yin Y, Ren H X, Liu X W(2008) Real-time wave simulation based on wave spectrum used in marine simulator. In: Proceedings of Asia Simulation Conference/ the 7th International Conference on System Simulation and Scientific Computing (ICSC'2008) ,2008,10:274–277

[32] Ren H X, Yin Y, Jin Y C(2008) Real-time shadow rendering in large-scale scene (In Chinese). Journal of Dalian Maritime University. 34(1): 58–60

[33] Ren H X, Yin Y, Jin Y C (2001) Scene modeling in navigation simulator (In Chinese). Journal of Dalian Maritime University 27(4): 20–23

[34] Li Y J, Jin Y C, Yin Y et al (2008) Simulation of shallow water wave in coastal region for marine simulato. In: ACM SIGGRAPH International Conference on Virtual-Reality Continuum and Its Applications in Industry (VRCAI 2008), 12

[35] Ren H X, Zhang X Y, Yin Y(2008) 3D real-time rendering of continuous oil spill on simulated ocean (In Chinese). Journal of System Simulation 20(19): 5117–5120

[36] Yin Y, Jin Y C, Ren H X (2003) Generation of optimal triangular meshes for simulation using data from contour lines (In Chinese). Journal of Dalian Maritime University 29(4): 14–17

[37] Yin Y, Ren H X, Jin Y C (2002) Graphic technology of distributed marine simulation system — V.Dragon 2000 (In Chinese). Journal of System Simulation 14(5):617–619

[38] Ren H X, Jin Y C, Yin Y (2004) Modeling of Underwater Scene in Simulation System (In Chinese). In: Proceedings of China Graphics Conference (Chinagraph' 2004), 2004,9:384–386

[39] Sun T D, Yin Y, Jin Y C (2005) Realization of binocular function in visual system of navigation simulator (In Chinese). Journal of System Simulation. 17(9) :2295–2297

[40] Zhang X Y, Yin Y, Yang X(2007) Research and application of stereo display in the visual system of ship handling simulator. In: Proceedings of Chinese Education Game and Virtual Reality Conference, August 2007

[41] Zhao M, Jin Y C, Yin Y(2008) The construction of CAVE stereo display system with geometric distortion correction (In Chinese). Journal of Computer-Aided Design and Computer Graphics 20(10): 1366–1373

[42] Ren H X, Jin Y C, Yong Y(2007) Real-time shadow rendering in scene system of marine simulator. In: Proceedings of the 4th International Conference on Image and Graphics (ICIG 2007), 2007, voll: 1010–1014

[43] Sun T D, Yin Y, Jin Y C (2002) Realization of viewpoint tracking in visual system of navigation simulator (In Chinese). Journal of Dalian Maritime University 28(1): 26–30

[44] Yin Y, Liu X W, Li Z H(2007) Key technologies of navigation radar simulator using real radar monitor (In Chinese). Journal of System Simulation. 19(5):1014–1017

[45] Ren H X, Jin Y C, Yin Y (2007) Rendering the sun and lens flare in visual system of simulator (In Chinese). Journal of Dalian Maritime University 33(2):124–128

[46] Yu Z D, Yin Y(2008) The design and development of UAIS simulator(In Chinese). Journal of System Simulation 20(1): 101–104

[47] Chen Y Z, Jin Y C, Liu X W(2004) Research on GPS Simulator based on GPS OEM (In Chinese). Journal of Systm Simulation. 16(8): 1754–1757

[48] Zhao M, Li S L (2006) System Simulation of Ship borne Inmarsat-B (In Chinese). Journal of System Simulation 18(1):67–72

[49] Wang Y Q, Bao X G (2006) Research on GMDSS module of full mission navigation simulator (In Chinese). Journal of Navigation Technology. June 2007:36–38

Author

Prof. Yong Yin was born in 1969. He received his B.S. and M.S. degree in Navigation Technology from Dalian Maritime University in 1991 and 1994 respectively. In 2001, he gained Ph.D in Transportation Information Engineering and Control from Dalian Maritime University. His research interests include navigation simulation technology, virtual reality and computer graphics in real-time. His doctoral thesis supervisor was Professor Yi-cheng JIN.

Since 2004, Prof. YIN has been a professor of Transportation Information Engineering & Control at the Key Lab. of Marine Simulation & Control of Communication Ministry, China (located on Dalian Maritime University). Up until now he has been an advisor for 5 doctoral and 14 postgraduate candidates. Since 2003, he has been a member of Chinese Navigation Society, System Simulation Society, Image and Graphics Society. In 2008, he was accepted as a committeeman of Virtual Reality Committee of Chinese Computer Society, Chinese Image and Graphics Society. Now he is also the director of Nautical Science and Technology Insititute, Dalian Maritime University. By now, he has written more than 70 papers about navigation simulation, application of virtual reality and computer graphics for scientific journals, as well as for national and international conferences.

Corresponding author

Yong Yin(✉)
Key Lab. of Marine Simulation & Control, Dalian Maritime University, Dalian 116026, China
E-mail: bushyin_dmu@263.net

Virtual Assembly Environment for Product Design Evaluation and Workplace Planning

Xiumin Fan[1,2], Rundang Yang[1], Dianliang Wu[1,2] and Dengzhe Ma[1,2]

1. CIM Research Institute, School of Mechanical Engineering, Shanghai Jiao Tong University

2. Shanghai Key Lab of Advanced Manufacturing Environment

Abstract

DFX technologies require evaluating the multiple aspects of product performance as early as possible, while virtual assembly with digital model can be used to realize the DFX idea. By using human-friendly oriented VR technology, the digital model evaluation for assembleability can be performed at the early stage of product development, so as to replace the evaluation of real physical model as much as possible. With the purple of building a virtual assembly experimental environment, system architecture of virtual assembly environment for product design and assembly workplace evaluation has been proposed, and key technologies for implementing virtual assembly environment have been discussed, such as VR application oriented general software development platform, virtual model descriptions, virtual assembly operation based on constraint and degree of freedom analysis, multiple interaction mode and ergonomic evaluation for user's assembly operation and so on. At the end, two kinds of applications for ship pipeline and auto engine have been carried out respectively to demonstrate the feasibility and effectiveness of virtual assembly environment technology.

Keywords

Design for Assembly, Virtual Reality, Virtual Assembly, Constraint and degree of freedom analysis, Ergonomic evaluation

1 Introduction

1.1 Trends of digital technology

For increasing the efficiency of product development and saving cost and time, the digitized degree of computer aided technologies for product development is continually improved; the trend is shown in Fig. 1. By directly designing the physical prototype manually, the product developers can obtain the most genuine multi-channels perception, however it is of high cost on time and spending. With the progress of technologies, it shows the development trend from the early 2D/3D computer-aid drawing tools to the current CAD system. Three dimensional systems can well describe the product's shape and structure. But for the goal of improving the designers'understanding of the product's genuine features, based on the 3D system, there appears system which can describe product physical attributes and perform product performance simulation, i.e. digital mockup simulation. With the advent of Virtual Reality (VR) technology, Virtual Prototyping (VP) and Virtual Manufacturing (VM) become the current focus area in digital technologies. Virtual prototyping and virtual manufacturing, which are supported under virtual environment, greatly improve designer's perceptive degree for computer model. Two aims can be achieved by using virtual reality technology: increase the degree of freedom for product development, and obtain the high level details of information. By this way, on the earlier stage of product development, it becomes easier to make forecast decision and optimize implementation scheme. The Digital Factory (DF), which highly integrates the information of production, control, logistic, cost, and marketing etc., supports virtual product development from the perspective of factory operations.

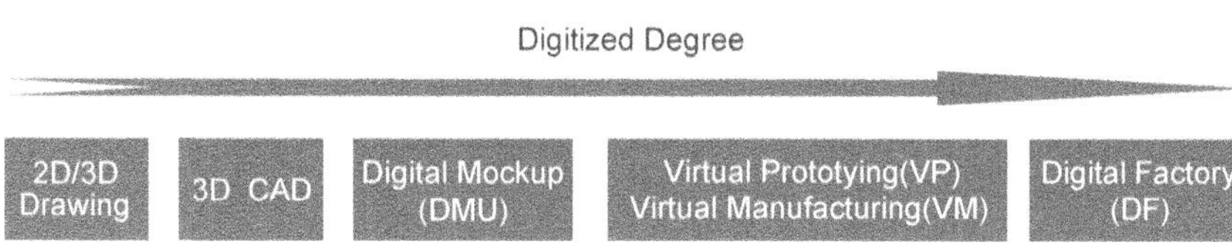

Fig. 1 The trend of digital technology

1.2 Background for building up virtual assembly environment

DFX technologies, which incarnate the idea of the concurrent engineering, require designers to evaluate the multiple aspects of product performance as early as possible, for example, to push forward the engineering design, manufacturing, maintenance and recycle tasks requirements as input to concept design stage. By using human-friendly oriented VR technology, the digital mockup evaluation could be performed at the early stage of product development, so as to replace the evaluation of real mockup as much as possible. Although currently there are many commercial 3D engineering tools that can be used for digital mockup (the number of tools is still increasing), there is still short of one key feature, the human-friendly nature and direct interaction with digital mockup. While the VR technology can provide these kinds of function and have its special advantages. The virtual testing environment, which is based on the virtual prototype and mapped to the real physical environment, provides an environment that doesn't consume any physical prototype cost.

1.3 Goal of building up virtual assembly environment

Virtual assembly environment must simulate the real physical assembly environment as much as possible. Through the trial assembly in virtual environment, the product developers can evaluate the components' assembleablility that includes reachable ability, visibility and difficulty of component replacement. For the aspect of workers' operation, the standardized and the least number of operation tools are expected, the reachable ability with assembly tool or by hand is necessary to be evaluated. In addition, this virtual trial environment can provide some assistant functions for assembleability analysis, such as assembly path computation that can be used for robot offline programming, the computation of moveable component sweep volume that can be used for consideration of internal reserved space, etc. By this way, the purpose is to replace the functions of physical assembly environment with virtual environment as much as possible, and to provide a decision support for engineers with different professional knowledge.

1.4 Three levels for assembly process oriented evaluation

There are three levels of method for digitized aid technologies that are used for product assembly process evaluation: the automatic assembly with fixed path, the assembly based on virtual human model and the assembly with real operators by using VR peripherals. The automatic assembly with fixed path (Fig. 2), which needs to define the assembly sequences in advance and shows the assembly process by exploded view, can only do evaluation based on vision information, and can only be browsed without nature interaction operations. The assembly based on virtual human model (Fig. 3), which needs to create the virtual assembly environment and pre-defines each action that the virtual human needs to conduct in the process of component assembly, can only do evaluation by the vision information and ergonomic analysis data. The assembly with real operators by using VR peripherals (Fig. 4) has the best advantage of initiative and interactivity. By

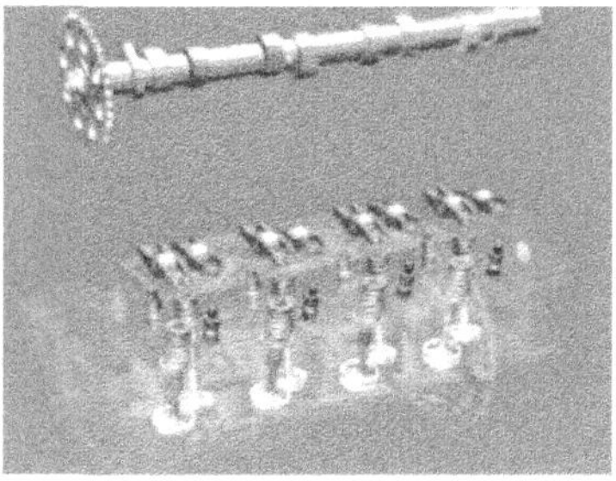

Fig. 2 Automatic assembly

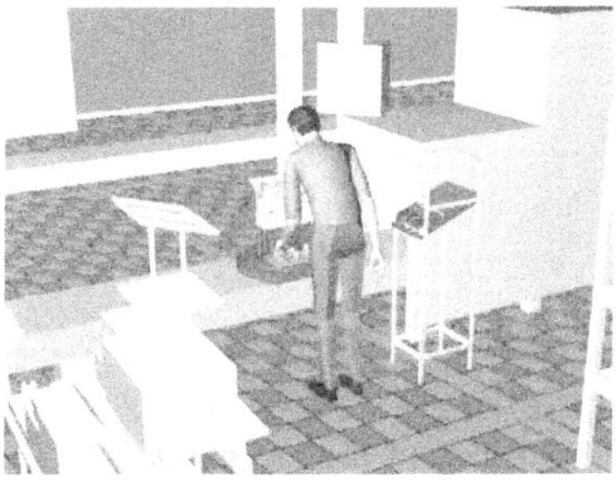

Fig. 3 Virtual human assembly

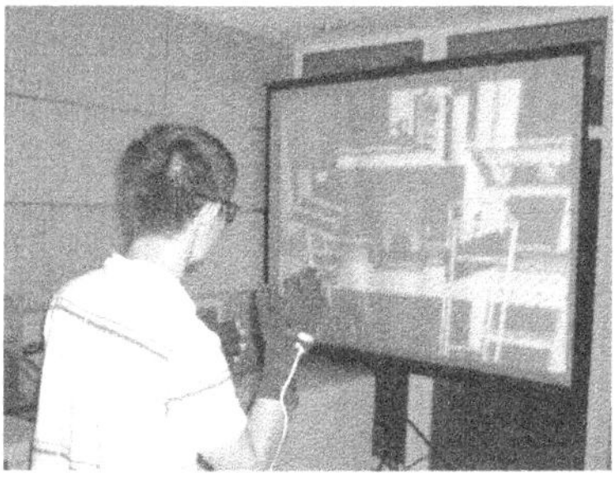

Fig. 4 Assembly with VR peripherals

this means, user can select assembly tools, pick up components, and perform the trial assembly under the virtual assembly environment. The evaluation of assembleability is not only based on vision information and ergonomic analysis data, but also the active participation from different roles of engineers.

2 System architecture of virtual assembly environment system

Figure 5 below shows the system architecture. The system includes several modules, such as the transition of product information from CAD, the low-level model library for product assembly, resource model library for assembly workplace planning, VR general development platform, etc. At the virtual assembly trial environment for product design, the operator can operate by manual, perform the analysis of assembleability and plan the assembly sequence. At the virtual assembly trial environment for assembly workplace planning, the operator can use assembly tools, perform the analysis of assebleability, record the assembly path, design the layout of workplace facilities, and do ergonomic evaluation. Through the multi aspects evaluations for the virtual trial assembly, the evaluation result can be fed back into CAD environment.

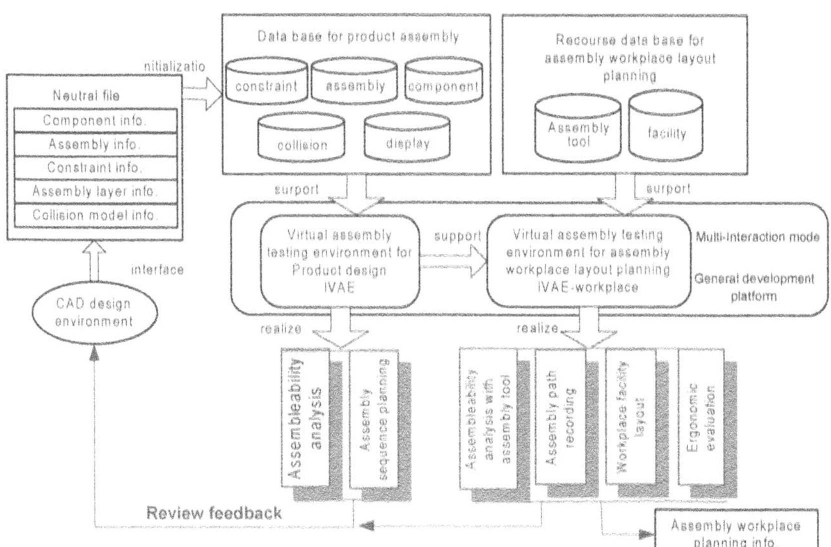

Fig. 5 System architecture

3 Key points for implementing virtual assembly environment

The generation of the virtual assembly trial environment involves many aspects of technologies [1–2], the key point include: the general software platform VR-Flier (developed by our Lab), the different model descriptions in virtual assembly environment, the virtual assembly operation based on constraint and degree of freedom analysis, the interaction methods in virtual assembly environment, and the Ergonomic evaluation for user's assembly operation, etc. The first key point is the general development platform.

3.1 VR application oriented general software development platform

The boom of VR application promotes the research and development at different fields. While at the same time, it also puts forward more and more strict requirements for developers. The development of VR applications not only involves the knowledge on different application fields, but also requires the knowledge of different implementation technologies for VR system, such as geometry modeling, visualization, stereo display, acoustics processing, sensor, telecommunication, etc. In general, the application developers mainly master the knowledge and technologies in their professional fields. For leverage the requirements for VR application developers, a "VR Application Oriented General Software Development Platform" has been built[3], which encapsulates bottom layer knowledge and technologies. This platform can run independently, at the same time, it is also an integrated general VR tools package that is easy to be used and extended, and it can support the re-development conveniently. The aim of this platform is to support the application development for different fields' "creation", to be used easily by not-specialized developers.

Figure 6 shows the hierarchy architecture for supporting the VR application system, it includes hardware layer, system software layer, tools library layer, general application layer and special application layer. As shown in Fig. 6, VR-Flier is belonged to general application layer that has already encapsulated the following layers — hardware layer, system software layer and tools library layer. It orients to the different special application. By this means, for different VR applications, the developers don't need

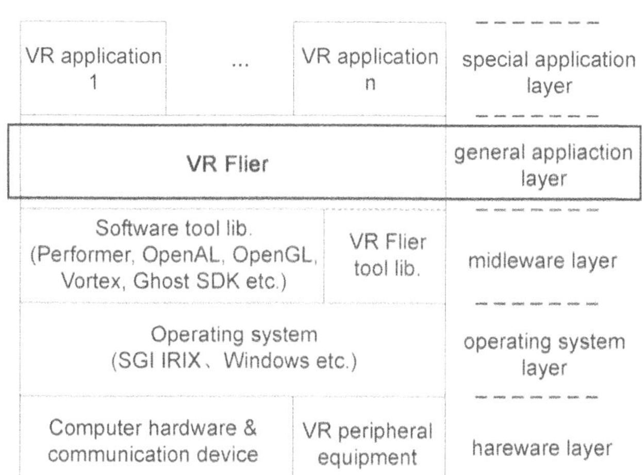

Fig. 6 General Software Platform Architecture for VR applications

to know the detail of bottom-level software and hardware.

3.2 Model descriptions in virtual assembly environment

The entities in virtual assembly environment include components, assembly tool and workplace facilities. The description of component includes geometrical information, design information, at the same time, the convenience of model organization and operation should be considered so as to meet the requirement of real time interaction. The geometrical information of assembly tool is described with B-rep method. The description model of assembly tool includes five kinds of attributes, i.e, query, physical, operation, geometry and operating object attributes. The description model of workplace describes information about storage rack, work bin, fixture, worktable and components related with this workplace.

In order to realize multiple function of entities operation in virtual environment, the organization of these entity objects is based on a set of models, i.e., the facet model for display, the collision model for collision diction and the logic model for describing component attribute [4–5]. The logic model is used to store the component's precise geometrical, topological and engineering information. The display model is used to render and display the component's facet model. The collision model is a kind of hierarchy convex model which is formed by facet model. The description of the above three kind of models is based on the same coordinate system, and these models are updated synchronously with position information from virtual hand, as shown in Fig. 7.

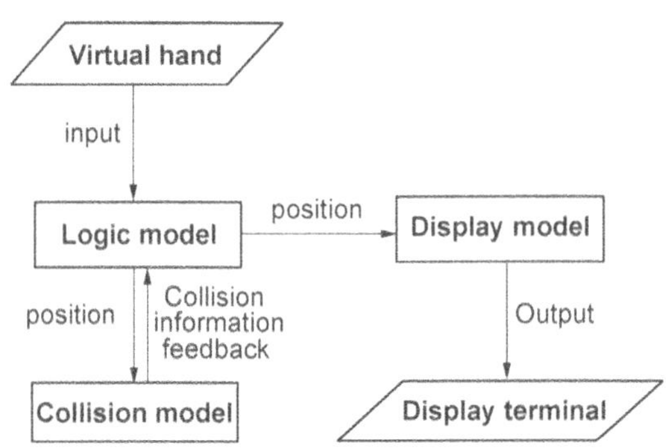

Fig. 7 Model organization of entities

The product's assembly tree model describes the relationship among different assembly entities, as shown in Fig. 8. Geometrical constraint model is a kind of very important information which is used to assemble component to form assembly and to set up the relationship between assembly tool and component. It is not only the primary method of precise positioning, but also the important approach to realize the assembly process. Eight kind of geometrical constraint have been considered, i.e., fit, fit offset, align, align offset, superposition, angle orientation, tangency and coordinate system.

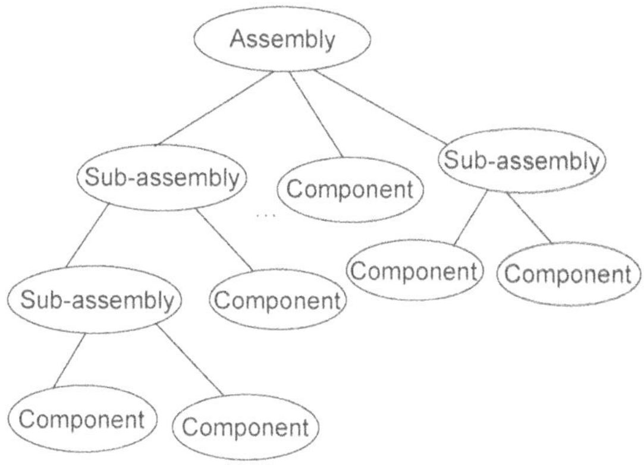

Fig. 8 Structure of product assembly tree

For the research on digital product assembly planning, the assembly sequence generation is the kernel concern; it is the precondition of assembly process plan evaluation and optimization, it is the basis of assembly facility selection, fixture selection and assembly operation simulation. Under this virtual assembly environment, the product assembling is completed by the interactive operation from operator; therefore, the whole assembling process implies the assembly knowledge and rule. By this way, the efficiency of the assembly sequence generation would be increased greatly. The assembly process not only contains the product assembly sequence, but also contains the description of assembly path and assembly track, as shown in Fig. 9–11. The assembly sequence includes sequence attribute, query attribute of sequence nodes, correlative entity attribute and the corresponding path. The assembly path includes path ID, geometrical attribute, animation attribute. The path track describes the information of control point and the track display attribute.

Fig. 9 Key point on assembly path

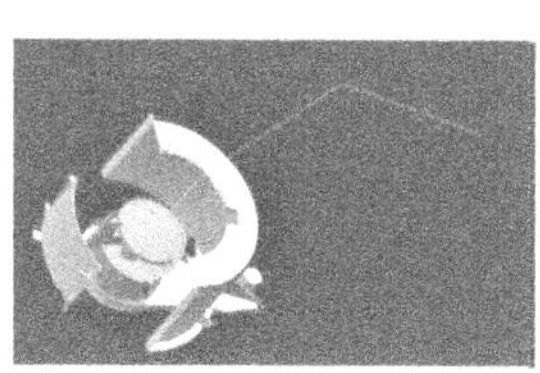

Fig. 10 Assembly path track line

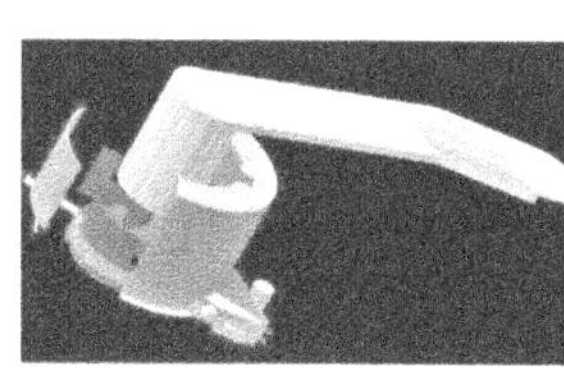

Fig. 11 Assembly path sweep volume

To describe the assembly workplace, the model would not only contain the entities' information and the relationship among these entities, but also the workplace operating information and the assembly operating information, therefore, a composite workplace description model is put forward which is assembling process and behavior oriented. This composite model describes contents as operation information on the workplace, information of component to be assembled, assembly tool and fixture, storage rack, guide rail and shop floor, as shown in Fig. 12. The assembly workplace model is the bridge for the organization of objects in virtual scene and contains the necessary information for workplace layout.

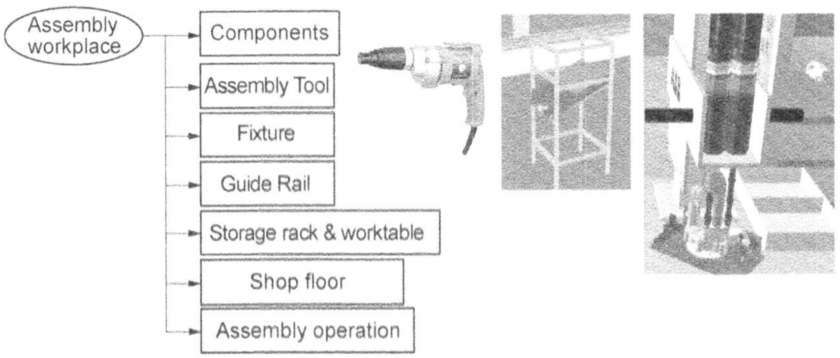

Fig. 12 Description of workplace

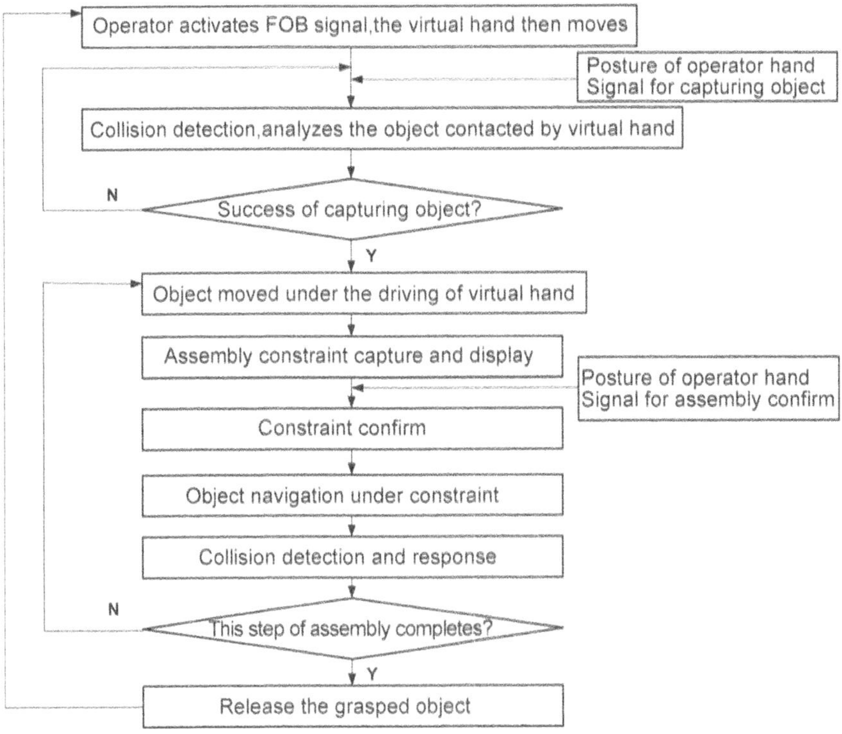

Fig. 13 The flow chart of basic virtual assembly processes

3.3 Virtual assembly operation based on constraint and degree of freedom analysis

As shown in Fig. 13, the flow chart of basic virtual assembly processes. The key steps are collision detection, constraint capture, confirm and navigation [6–7]. Virtual assembly operation based on constrains and degree of freedom (DOF) analysis is a very important approach to realize the component's precise positioning. The purpose of DOF analysis is to set up equivalent motion DOF for each type of constraint based on the component's assembly geometrical constraint, and to obtain movable space for component being operated by DOF consolidation. The step of virtual assembly based on constraint and DOF includes uniform description for geometrical constraints, setting up the equivalent relation between constraints and DOF, normalization of motion DOF, algorithm for automatically assembly relationship recognition based on constraint, algorithm for component positioning computation and motion navigation.

The principle of constraint capture algorithm is to compute the current position of constraint related with this step of assembly operation, and then compares the constraint's parameters so as to make judgment whether the constrain could be captured or not. As shown in Fig. 14, an axis align constraint is captured and the axis of these two components is displayed to indicate the existence of this axis align constraint. Constraint confirmation is realized by command sent from data glove during the interaction process, as shown in Fig. 15, the relative position between the active object and passive object meets the constraint checking requirement. Constraint confirmation is in fact the constraint's position computation. It means to satisfy the new constraint with updated position adjust value without breaking other exiting constraint already confirmed. Constraint navigation is very important for component's precise positioning and movement navigation during the assembly operation processes. As shown in Fig. 16, the grasped component could only move along the axis. In the virtual assembly environment without force feedback, the function of constraint for component positioning and assembling navigation is much more significant.

Fig. 14 Constraint capture

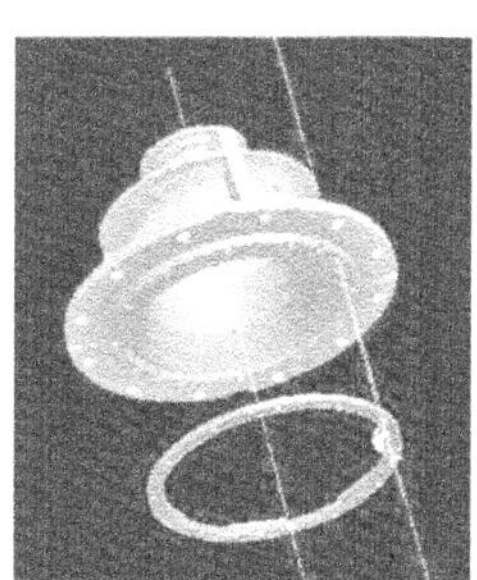

Fig. 15 Constraint confirmation

Fig. 16 Constraint navigation

The hierarchy convex model is generated during the initialization stage and used for interference computation during collision detection. Some other speedup algorithms are also adopted, for example, in case that model pairs for collision detection need to be decided dynamically. By this means, the number of models for collision detection could be decreased greatly. In the virtual assembly environment for assembly workplace layout planning, collision detection is not only needed between facility objects, components and shop floor surrounding environment, but also needed for component grasped and failing down to the floor. As shown in Fig. 17, the collided component is highlighted when collision is detected.

According to different operating status when object is moved, collision response could be classified into three categories, i.e. collision response between moving virtual hand and object, between objects, between operated object and ground when the object failing down. As shown in Fig. 18, during the workplace layout planning, the moving virtual hand collides with and grasps the storage rack; the result of collision response is that all the components on the storage rack could be move together with it.

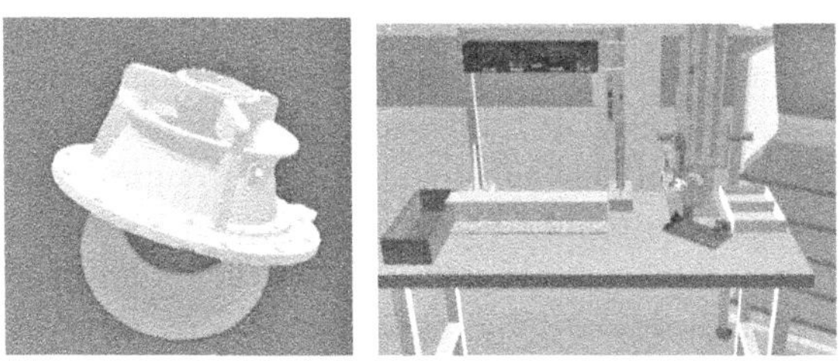

Fig. 17 Collision detection (Collided component)

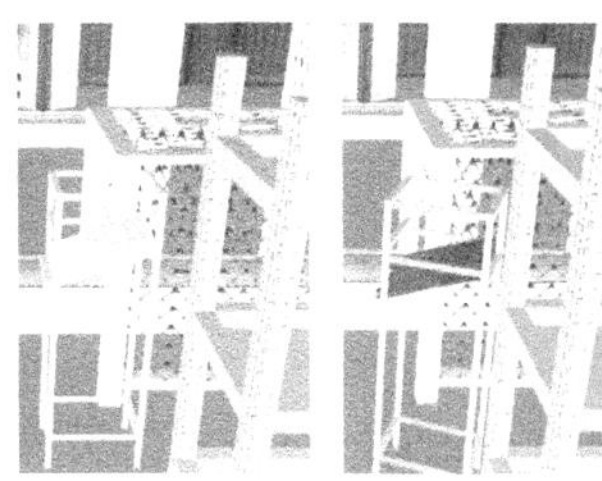

Fig. 18 Collision response

3.4 Multiple interaction mode in virtual assembly environment

The running flow chart of virtual assembly system is shown in Fig. 19. Besides virtual hand interaction with data glove, there are other kinds of auxiliary interaction mode, such as speech, virtual menu, dialogue etc. At different kind of assembly stage, user could use different kind of interaction mode, so as to increase the user's initiative and the assembly efficiency.

For the interaction mode based on virtual hand, the data glove and position tracker (flock of bird, FOB) are mounted on user's hand which is corresponding to the virtual hand. User can control the virtual hand to realize the assembly intention, as shown in Fig. 20. For the walk-through mode in virtual scenes, there are two approaches. One is by FOB which is mounted on user's head; the walk-through is realized according to the view point information updated by user's head movement. Another one is by Neowand (a device with small patch of button), the view point is changed by pressing the up, down, left and right button.

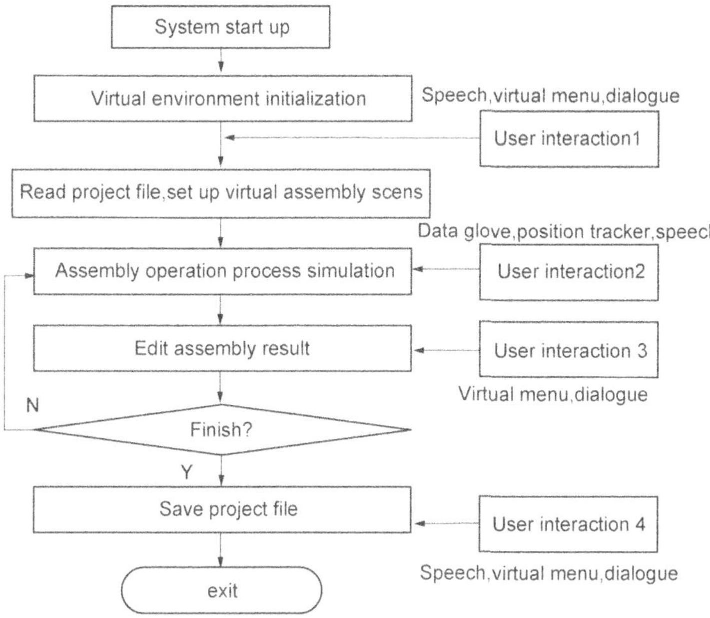

Fig. 19 The running flow chart of virtual assembly system

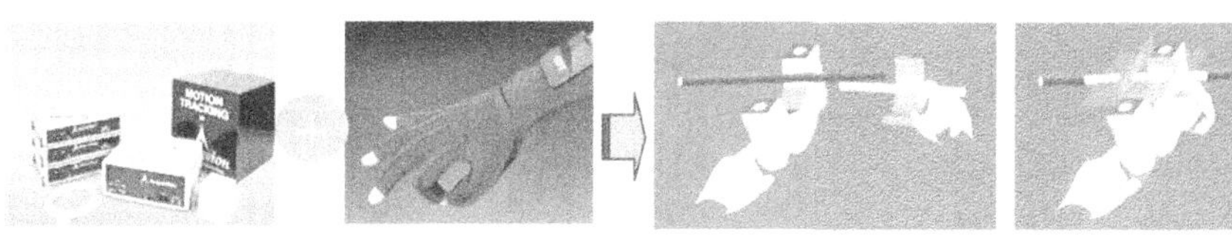

Fig. 20 Interactive assembly based on virtual hand

3.5 Ergonomic evaluation for user's assembly operation

For the evaluation of assembly workplace based on ergonomic analysis, it is necessary to gather the user's position data through position tracker. A scenario has been designed to gather data, as shown in Fig. 21, FOBs are put on user's back and hands separately.

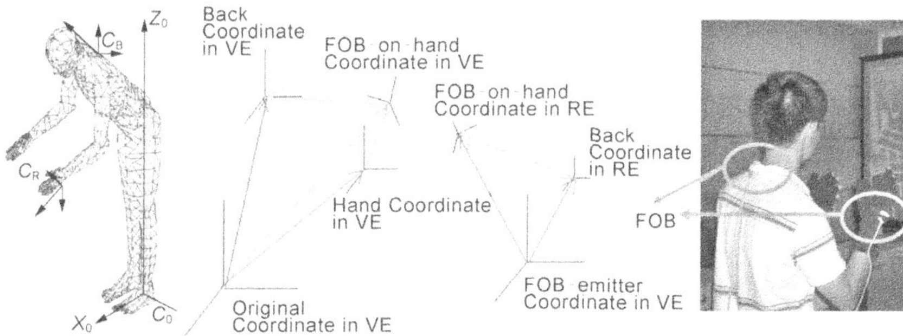

Fig. 21 Mapping relation between virtual hand and real hand

As the virtual hand is drove by relative incremental movement of user's hand, from the point of view of world coordinate system, the virtual hand is not strictly consistent with the real hand. Therefore the mapping relation between virtual hand and real hand should be setup, by this means; the corresponding position of the real hand in the virtual environment could be calculated.

Based on the position information of user's body gathered during the assembly procedure, geometrical parameters could be obtained as shown in Table 1, such as the hand's position, vertical distance, degree of asymmetry, horizontal distance and distance of object being lifted up or put down. According to these data, the assembly workplace layout planning could be evaluated through ergonomic analysis. The indexes used for evaluation are user's reach ability, field of view, maximum force and torque exert on user, posture analysis, NIOSH fatigue analysis, time of fatigue & recovery and energy consumption etc. The two indexes used here are NIOSH evaluation and energy consumption. For the first one, the equation $CLI=STLI_1+\sum \Delta LI$ is used to calculate the index value; for the second one, the equation $A=\left(F\cdot H_n+\frac{F\cdot L}{9}+\frac{F\cdot H_0}{2}\right)\cdot K\cdot n$ is used. Only these two values are within the standard arrange, the workplace layout could be said satisfied.

Table 1 Data sheet for measurement

Item	Object Weight (Kg)		Position of hand(cm)				Vertical distance (cm)	Asymmetry (degree)		frequency	cycle	Coupling coefficient
			Start		end			start	end	time/minute	hour	
	Lav	Lm	H	V	H	V	D	A	A	F		C
1	22	33	16	0	16	30	30	0	0	1	8	Fair
2	33	44	12	0	12	6	6	0	0	2	8	Fair
3	11	22	8	30	8	39	9	0	0	5	8	Fair

4 Applications

Based on this virtual testing environment for product design and assembly workplace layout planning, two kinds of applications have been carried out. The first one is about the ship pipeline system assembly. As the pipeline system is very complex, during the pipeline system assembly of real ship, rework is occurred from time to time, this is the reason why it is idea to do pipeline assembly in this virtual environment. Figure 22 shows the ship body, and Figure 23 shows the assembly scene of pipe component, valve, flange and other accessories. By this means, the feasible assembly sequence and assembly path can be obtained. To setup the virtual prototype for ship is of very important significance, because potential mistake could be avoided in the early design stage, so as to reduce rework in field, reduce cost and ship development cycle time.

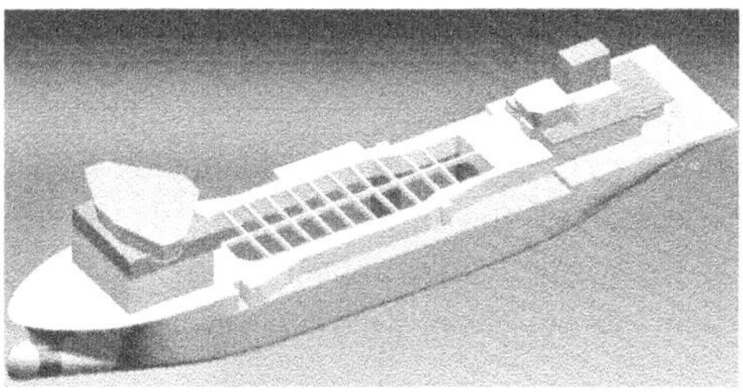

Fig. 22 Ship body

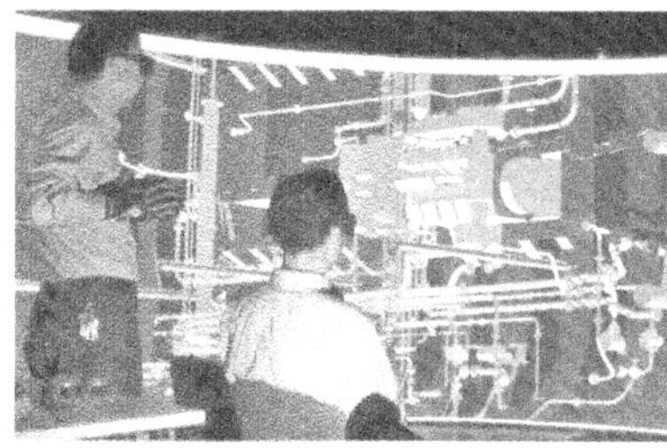

Fig. 23 Assembly of pipeline system

Another application is about engine assembly workplace planning. To meet product throughput requirement, the total assembly time for each assembly workplace in the assembly line should meet cycle time requirement. While the assembly operation time has close relation with the workplace facility layout, the assembly operation with assembly tool etc. Therefore it is necessary to have a virtual testing environment to do evaluation first. Figure 24 shows workplace layout for crank shaft assembly. Firstly, based on the process plan, workplace facilities, assembly tool and fixture can be selected from data base. Then components can be assembled one by one. Finally, ergonomic evaluation can be conducted. According to interactive assembly processes, the assembly process plan can be checked and the workplace layout plan can be obtained.

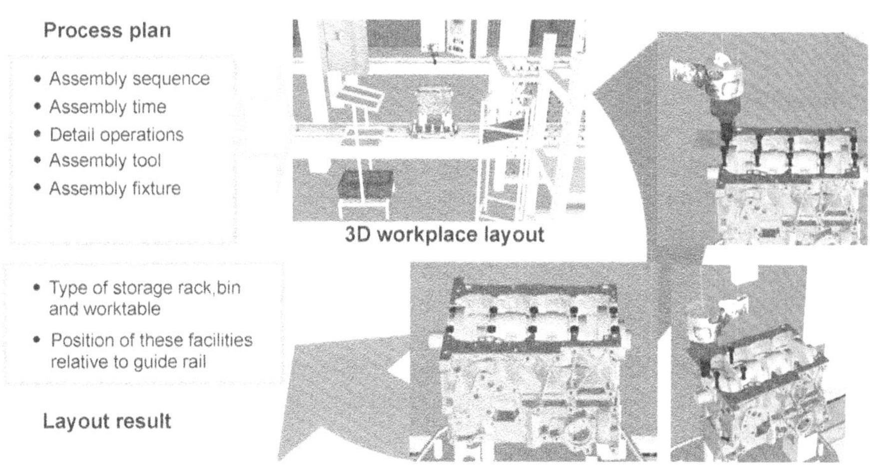

Fig. 24 Assembly workplace planning and assembly process evaluation

5 Future Works

One important research direction is virtual assembly with haptic feedback. It can provide haptic perception for user. By this means user can sense the component's physical attribute during the assembly procedure. It needs to build the haptic calculation model and the haptic optimization model which considers all five fingers grasping object with random shape. And it also needs to plan the haptic distribution among fingers and palm. So the haptic generation and feedback system can be set up, with this user can fell the grasped object.

The development of complex product is usually involved lots of component supplier who locate at different place. In order to make sure the product quality, communication between each other is necessary during the product development process. Nowadays computer network technology has made the internet based collaborative work possible. To work collaboratively by organizing resource located at different place, it is an area paid attention by researchers. In collaborative virtual assembly environment, components from different suppliers could be assembled into product conveniently, so that the assembly of product digital model could be checked and simulated at early design stage.

Acknowledgments

This work is supported by Key Project from Science & Technology Commission of Shanghai Municipality (Grant no. 08DZ1121000) and National High-Tech Program of China (Grant No. 2006AA04Z141).

References

[1] Yang R D, Wu D L, Deng H L, Fan X M, Yan J Q (2004) Research and implementation of product assembly technology in virtual environment (in Chinese). Computer Integrated Manufacturing Systems 10(10):1220–1224

[2] Wu D L, Yang R D, Ma D Z, Fan X M (2004) Virtual Assembly System and the Enabling Technologies (in Chinese). Journal of Shanghai Jiaotong University 38(9):1539–1543

[3] Pi X Z, Fan X M, Yan J Q (2005) VR Flier:A Software Platform for Virtual Reality General Application Development (in Chinese). Journal of System Simulation 17(5):1157–1162

[4] Wu D L, Yang R D, Ma D Z, Fan X M (2004) Product information representation in integrated virtual assembly environment (in Chinese). Computer Integrated Manufacturing Systems 10(11):1364–1369

[5] Zhu W H, Fan X M, Ma D Z (2005) Research on data transformation from CAD system to integrated virtual assembly environment (in Chinese). Computer Integrated Manufacturing Systems 11(8):1115–1119

[6] Yang R D, Wu D L, Fan X M, Yan J Q (2006) Research on constraint-based virtual assembly technologies (in Chinese). Computer Integrated Manufacturing Systems 12(3):413–419

[7] Wu D L, Yang R D, Ma D Z, Fan X M (2004) Constraints Navigation in Integrated Virtual Assembly Environment (in Chinese). Chinese Journal of Mechanical Engineering 40(11): 47–52

Author

Xiumin Fan is a professor at Computer Integrated Manufacturing Research Institute, School of Mechanical Engineering, Shanghai Jiao Tong University. During the study period from 1989 to 1998 in Shanghai JiaoTong University, she received Doctor Degree (1998) in Mechanical Engineering Department. In 2004, she received Shanghai Women Innovation Nomination Award. In 2006, she received the 10th Ying Tung Fok Education Foundation Award for Young Teachers (research category). In 2007, she was selected to joint the Program for New Century Excellent Talents in University from The Ministry of Education. Her research interests are virtual prototyping and virtual manufacturing.

Rundang Yang is a Postdoctoral Fellow at Shanghai Jiao Tong University. His research interests are virtual assembly and ergonomic.

Dianliang Wu is a lecturer of Shanghai Jiao Tong University whose research interests are virtual prototyping, virtual assembly, reverse engineering rapid prototyping and computer aided engineering. From 2003 to 2005, he was a post doctor fellow in SJTU. He received Ph.D. in Mechanical Engineering of Xi'an Jiao Tong University in 2003.

Dengzhe Ma is a professor of Shanghai Jiao Tong University (SJTU), the director of Shanghai Key Lab of Advance Manufacturing Environment and the previous head of Computer Integrated Manufacturing Research Institute of SJTU. He received Bachelor Degree in Mechanical Engineering from SJTU in 1964 and his Dr.-Ing. Title in Mechanical Engineering Department of Darmstadt Technology University in Germany in 1987.

Corresponding author

Xiumin Fan(✉)
CIM Research Institute, School of Mechanical Engineering, Shanghai Jiao Tong University, No.800, Dongchuan Road, Shanghai 200240, China.
E-mail: xmfan @sjtu.edu.cn

Numerically Controlled Virtual Models for Commissioning, Testing and Training

Marco Schumann, Michael Schenk and Eberhard Bluemel
Fraunhofer Institute for Factory Operation and Automation, Germany

Abstract

The increasing complexity of machines is demanding significantly more time and effort to commission them. At the Fraunhofer IFF Virtual Development and Training Centre VDTC in Magdeburg, components of a machine control system can be tested on a virtual model of a machine before the real machine goes into operation.

Keywords

Virtual Reality, Virtual Commissioning, Virtual Training

1 Motivation

The manufacture of automated manufacturing equipment such as machine tools or special machines is characterized by intense cost, time pressure and increasing demands on product quality. The time and effort spent in developing and manufacturing new products are evolving into a crucial competitive factor. Not only product development but also service, training and marketing are increasingly playing a role.

For the most part, automated manufacturing equipment is developed sequentially. The steps of mechanical design engineering, electrical design engineering and control system development are completed successively. Control software in particular is developed and first tested on the real machine. The increasing complexity of manufacturing equipment is causing problems ever more frequently. Mistakes from earlier stages of development often only become apparent when a machine is commissioned. This generates further iterative cycles in development and prolongs the time until a machine is commissioned. In the worst case, this is connected with exploding costs. A sound test of the software on the real machine is often impossible for reasons of time.

2 Virtual Models in the Product Life Cycle

Interactive visualizations and simulations enable using a computer to clearly and realistically reproduce complex structures and processes. Virtual models can already convey a comprehensive, three-dimensional impression of a product in early phases of development. The preliminary design can be tested, dimensions can be checked and potential error sources can be identified before implementation. Consequently, a key field of application is the use of digital models for the integrated planning, validation and control of product development processes [1, 2].

Moreover, interactive visualization and simulation are increasingly growing in importance for sales since interested parties can be given a convincing impression of a finished machine or a complete system even before making a purchase. In the planning and design phase, a machinery manufacturer's client can be involved in the broader development and production process so that the product produced is customized for a client's specific needs.

3 Creating Virtual Models from Existing Data

The time and effort that go into creating virtual models and incorporating them in existing VR systems is still relatively substantial though. Specialized software solutions exist for many applications, which predominantly employ their own data formats. Therefore, our research work is focused on creating virtual models from electronically existing information. Typical solutions that are already operable include data transfer from CAD systems into interactive visualization systems [3], automatic generation of mechanical models from CAD data [4, 5] and the generation of electrical model from EPLAN data [6].

Frequent data conversions into different formats are laborious, may be afflicted by data loss and often only able to function in one direction. A 3-D CAD system and VR systems interface is a typical example. A CAD system operates with parameterized volume data for example. However, the overwhelming majority of VR systems employ polygonal surface models for visualization. Complex CAD data can often only be transferred into a VR system after great simplification. Modifications of the data in the VR system during a design review often cannot be returned to a CAD system at all and when they can then only with much work.

Over several years of development work, the Fraunhofer IFF produced a tool that allows transferring CAD data from commercially available systems such as Pro/Engineer, CATIA or SolidWorks to a VR system also developed by the Fraunhofer IFF. Correct object names as well as the assembly structure configured by a design engineer are transferred. Coordinated modules support potentially necessary reworking steps such as preparing different levels of detail, smoothing edges or applying textures. This minimizes the work required to generate the geometry model.

The geometry model serves as the foundation to create the functional model, i.e. the mapping of a machine or plant's performance. An object's performance in a virtual model is described in a completely different programming language, for example, a CNC machine tool's motional sequences. If the virtual model is intended to be used for training purposes, it is necessary to elaborately convert the real control components (CNC or SPS programs for instance) into the programming code for the VR system. The required work is so great in practice that it outweighs the potential benefits of VR supported training systems. Thus, this is one of the reasons why only a relatively low number of VR based training systems are in use so far.

Here, the only feasible solution is to connect the virtual model with the controller software or hardware of the real machine. This approach is described in the next sections.

4 Functional Tests on the Virtual Model

The commissioning phase does not act as a factor prolonging the overall development time of equipment. The ability to perform functional tests at a time as early as possible is advantageous. In practice, a functional test can only be performed when the design engineering has been completed. Only when a virtual prototype is used can several development tasks proceed simultaneously (simultaneous engineering). The functional test can already be started on digital models in the virtual environment while the machine is still being manufactured.

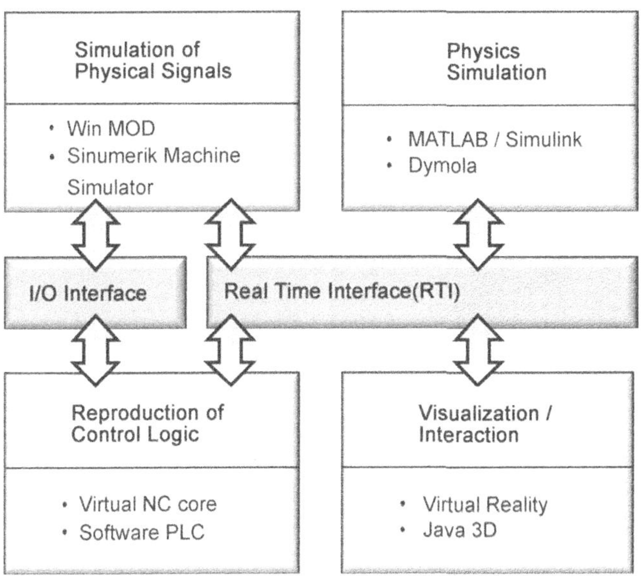

Fig. 1 Using the run-time interface to connect different components of the virtual machine

The success of the virtual environment essentially depends on the extent of how easy it is to include data from existing virtual engineering tools. These tools can be classified into the following groups:

- **Reproduction of control logic:** The most realistic approach to replicate the behaviour of a real machine is to use the identical machine control algorithms. In general, there are two alternatives for coupling the machine control algorithms with the virtual model. It can either be accomplished by using the real hardware (CNC control unit or programmable logic controller, PLC) or via integrating software (virtual numerical control core or PLC software) in the virtual model [6].
- **Simulation of physical signals:** In case the real control hardware or software is used it is also required to synthetically generate the according sensor signals. For this purpose a variety of simulation software exists. Examples include the Sinumerik Machine Simulator [7] or Winmod [8].
- **Physics simulation:** In some cases, the generation of the sensor signals can be rather complex — especially if physical correctness is required. In this case it might be useful to execute these calculations in mathematical tools, such as Dymola or MATLAB/Simulink and establish a real-time link to the virtual model [9].
- **Visualization/Interaction:** There is a wide variety of visualization software available. Depending on the quality of the image the range of tools extends from the relatively simply VRML of X3D viewers to highly immersive virtual reality environment using multi display systems [10].

The different classes of tools and typical examples are illustrated in Fig. 1. The real-time communication between them is established via real-time interface [11].

Depending on the machine functions that are to be tested different combinations of these tools can be selected. The correct operation of the machine specific CNC cyles (e.g. for changing tools) can simply be verified by using a real CNC controller and the visualization (left part of Fig. 2). By adding a simulation of physical signals it is possible to test critical situations. Including a collision detection and material removal simulation allows for testing the CNC code (right part of Fig. 2).

Fig. 2 Testing the cycles for replacing the machine tool (left) and simulation of the material removal (right)

5 Connecting a Real Control System with the Virtual Model

In order to implement a close-to-reality testing environment we coupled the fully functioning control system of a heavy machine tool with its virtual model. Connecting the real control system with the virtual model world can parallelize the development process in the technical domains involved.

Design engineers work in their accustomed CAD environment. The control systems engineers are enabled to develop their software in parallel on the virtual model of the machine. They too can already work on the real control system. The virtual model of the machine created to do this consists of the kinematic model and the performance model. It reproduces the fundamental characteristics of the machine's performance. The machine's operating performance and disturbance response can already be tested during the development phase. Thus a machine can be optimized throughout the whole development process.

6 Operator Training in the Virtual Environment

Such a system can be used beyond the development process for many and diverse applications, e.g. operator training or CNC programmer training. The virtual machine model guarantees realistic machine performance. This makes training conducted on the machine especially demonstrative.

Fig. 3 Future operators are trained using the real control unit and a virtual model

Thus, manufacturers of technical assets can already provide their customers virtual training environments, while the real machine is still being manufactured. Customers have the advantage of being able to train their operators at an early stage in an environment in which potential operating errors cannot cause damage on a real machine. This additionally saves valuable time. Operators have already acquired initial experience handling a machine before it is put into operation.

Connecting the virtual model with the real control system during the machine's operation generates other advantages. On the one hand, model parameters can be obtained from the real machine's performance. This is used to render the machine model more precisely so that modifications of machine configuration can be tested and implemented parallel to ongoing operation. Thus, setup times can be minimized. Likewise, by coupling the machine model to the real machine, current operating parameters can be documented parallel to machine operation and the machine's condition can thus be diagnosed.

7 Virtual-Interactive Product Presentation Supports Sales

Visual-interactive reproductions of machinery and plants can significantly contribute to efficient product presentation and constitute a universal marketing instrument. Apart from their demonstrativeness, even including images and videos too, interactivity is another extremely important feature of product presentations that use virtual methods. Users can acquire their first immediate experience with a product and freely move in the virtual environment to explore a machine. It is possible to examine every assembly and every single element of a machine, depending on the representation's level of detail. Techniques that make certain machine components transparent or hide them also make it possible to view internal or difficult elements in the virtual world. If functional models underlie the machine's assemblies, it is possible to move a machine's individual parts and learn their functionalities. Naturally, this allows assembling and operating complete products, machines and plants.

Thus, virtual-interactive representations can effectively present a product's advantages as well as its design and operation. A comprehensively designed product scenario can be used for the widest variety of target groups, e.g. sales staff, operator personnel or service staff.

References

[1] Danneck U, Issler T (2008) Virtuelle Welten in der Automobilentwicklung; In CAD-CAM Report Nr. Dec. 2008; Dressler Verlag, Darmstadt

[2] Döbler T (2008) Simulation und Visualisierung in der Produktentwicklung; In FAZIT Schriftenreihe, Band 12

[3] Bluemel E, Hintze A, Schumann M, Schulz T, Stuering S (2003) Virtual Environments for the Training of Maintenance and Service Tasks, Proceedings of the Winter Simulation Conference, 7–10 December, 2003, New Orleans, LA /

USA
[4] Engelson V, Bunus P, Pipescu L, Fritzson P (2003) Mechanical CAD with Multibody Dynamic Analysis Based on Modelica Simulation; In Proceedings of the 44th Scandinavian Conference on Simulation and Modeling SIMS 2003, 18–19 September, 2003, Västerås, Sweden
[5] Engelson V (2000) Tools for Design, Interactive Simulation, and Visualization of Object-Oriented Models in Scientific Computing. Ph.D. Thesis, IDA, Linköping University, Sweden
[6] Böhme T, Kennel M, Schumann M, Winge A (2009) Automatisierte Erstellung domänenübergreifender Modelle und echtzeitfähige Kopplung von Simulation, Visualisierung und realen Steuerungen In: Augmented & Virtual Reality in der Produktentstehung, 8. Paderborner Workshop in der Produktentstehung, Nr.252, June 2009, HNI Verlagsschriftenreihe, Paderborn, Seiten 155–170
[7] Stoeppler G, Menzel T, Douglas S (2005) Hardware-in-the-loop simulation of machine tools and manufacturing systems, In: Computing & Control Engineering Journal, Vol 16, Issue: 1, pp 10–15, ISSN: 0956–3385
[8] Bathelt J, Meile J (2007) Computer Aided Methods Supporting Concurrent Engineering when designing Mechatronic Systems Controlled by a PLC, Proceedings of International Conference on Manufacturing Automation, 2007, Singapore, Singapur
[9] Juhász T, Schmucker U (2008) Automatic Model Conversion to Modelica for Dymola-based Mechatronic Simulation;In Modelica 2008 Proceedings, 6th International Modelica Conference, Modelica 2008, 3–4 March, 2008 University of Applied Sciences Bielefeld, Germany
[10] Schumann, M, Böhme T, Otto M (2007) Kopplung von CNC-Steuerung und virtuellem Modell, In: Schulze T, Preim B, Schumann H (eds) "Simulation und Visualisierung 2007", Proceedings der Tagung "Simulation und Visualisierung 2007" am Institut für Simulation und Graphik der Otto-von-Guericke-Universität Magdeburg am 8. und 9. März 2007, ASIM SCS Publishing House e.V., Erlangen, 2007
[11] Kennel M, Bayrhammer E (2007) Eine Schnittstelle zur echtzeitfähigen Kopplung hete-rogener Simulations-, Steuerungs-, und Visualisierungsapplikationen; In Forschung vernetzen — Innovationen beschleunigen 3./4. IFF-Kolloqium; 20. April und 28. September 2007, Magdeburg; ISBN: 978–3–8167–7557–7

Authors

Prof. Dr.-Ing. Dr. E.h. Michael Schenk was born in 1953; majored in mathematics at the Technical College Magdeburg; 1983, doctorate; 1988, post-doctoral degree from the Technical University Magdeburg; since 1994, professor for "Factory Planning and Logistics"; since 1999, professor for "Logistics Networks" at the Otto von Guericke University Magdeburg; since 1994, Director of the Fraunhofer Institute for Factory Operation and Automation IFF in Magdeburg.

Dr. rer.nat. Eberhard Bluemel is the Head of the Department Virtual Interactive Training at the Fraunhofer Institute in Magdeburg, Germany. He holds a Ph.D. in natural science from the University of Magdeburg. His research fields include discrete optimization, operations research, logistics, simulation, and Virtual Reality.

Marco Schumann is an employee at the Fraunhofer Institute in Magdeburg, Germany. He holds a Master's degree in Computer Science from the Otto-von-Guericke-Universität Magdeburg. His experiences include a one-year-stay at the University of Wisconsin — Stevens Point. His main research interest lies in application of simulation methods in the field of factory planning and distributed virtual training environments.

Corresponding author

Michael Schenk(✉)
Fraunhofer Institute for Factory Operation and Automation, Sandtorstrasse 22, 39106 Magdeburg, Germany
E-Mail: michael.schenk@iff.fraunhofer.de

Virtual Reality Boosting Automotive Development

Mangqin Jiang
Design Department, Pan Asia Technical Automotive Center (PATAC), China

Abstract

Several Virtual Reality (VR) centers have been built in the automotive companies in China since 2004. This article analysed the current status, achievement and issues in these VR centers, and described how to make better use of the VR technology, from the view of VR recognition, VR process and VR application. Then it explored how to improve the hardware and software of VR to meet the requirements of research and development of automotive industry so that the virtual reality could play the even more significant role in boosting the research and development of automotive products.

Keywords

Virtual Reality, Virtual Engineering, Virtual Prototype, Virtual Test, Virtual Manufacture

1 Virtual Engineering vs. Virtual Reality

Virtual engineering is becoming a popular word in Chinese automotive industry. A lot of Chinese motor enterprises are devoting to developing their virtual engineering capability. In October, 2007, SAIC held a forum and invited several famous experts to take part in it, who had offered their valuable suggestions to SAIC theme as system construction of capacity development, virtual engineering and advanced engineering. In the development scheme for 2009, Pan Asia Technical Automotive Center has regarded VE as the critical capability development trend in the future.

However, the virtual engineering actually is not the same as virtual reality. Nevertheless, in terms of current VE application in Chinese automotive industry, the trend is appearing to mix up VR and other computer-aided technology such as CAE analysis or simulation analysis; whereas sometimes VR is mistakenly regarded as the post-process for presenting CAE software, which is blocking the application of VE to certain degree. To fully recognize the value of VR and make better use of it, it is essential to clarify them.

First, let's take a look at the definition of VE. You may find thousands of VE definitions. From Wiki, virtual engineering is defined as integrating geometric models and related engineering tools such as analysis, simulation, optimization, and decision making tools, within a computer-generated environment that facilitates multidisciplinary collaborative product development. Virtual engineering shares many characteristics with software engineering, such as the ability to obtain many different results through different implementations [1].

Figure 1 shows connotative meaning of VE. VE integrates CAX technology, to complete the virtual prototype, virtual test and virtual manufacture, etc. with the support of multi-means such as virtual reality in entire environment supplied by PDM system.

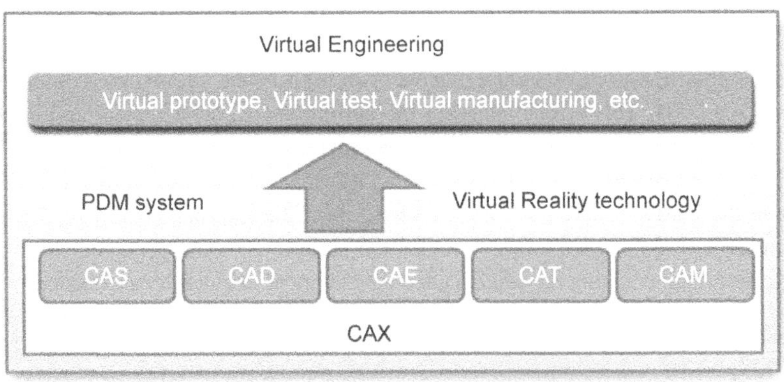

Fig. 1 Connotative meaning of virtual engineering

Now let's take a look at the definition of VR. Virtual reality was firstly raised up in USA in early 1980s, which means building a virtual world in computer and simulate the seeing, hearing and feeling to make the user immersed. Then the user could be interactive with the virtual objects in real time with no limits. VR has four features, namely multi-sensory, immersion, interactivity and imagination, which could distinguish VR from CAX. VR provides the more direct, simple, safe and interactive environment and adds the subjective feelings into the process, so that the digital property can be understood more clearly and thoroughly. Automotive engineers can review the design from any point of view in the virtual world where they could inspect the detail of parts, navigate in the car, even drive the car to evaluate the noise. What's more, VR can integrate the knowledge of multi-disciplines enabling people with different background to share[2].

According to the definition above, we can understand that VE is not VR. We should study VR deeply and make better use of it.

2 Widely Application of Virtual Reality Technology in Automotive Companies over the World

At present, VR is widely used in the automotive industry in USA, Japan and Europe etc. Almost all the main automotive enterprises, such as GM, Ford, Chrysler, Audi, Benz, Porsche, BMW, Volkswagen, Honda, and Toyota,are widely using the VR and scheme. Doubtlessly, the VR shall bring great profits for automotive companies, which has been widely recognized. The VR centres have been established one by one by the automotive companies in the world. GM's VR centre is an early one launched around 1997, equipped with Powerwall and CAVE for reviewing design, evaluating the view of driver and simulating the assembly of the parts. In 1998, Benz completed its digital prototype with VR technology. In April 2000, Benz spent 8 million Marks in building its VR centre. In 2004, Volkswagon spent 20 million EU$ in establishing two visualization centres—one is a powerwall for design review and the other is a CAVE for ergonomics, driver view and interior evaluation and drive simulation. Furthermore, other motor companies have also built their VR centres or actively applied VR in the development. For example, in Mazda, the customer can wear an HMD and groove to change the virtual car's color and accessories, and choose their favourite cars. In Ford, powertrain engineer simulates the engine maintenance in VR. In Nissan, the product line is simulated. The car design companies, such as ItalDesign, Pinifarina,have their VR centres as well. Recently, more and more VR centres are built all over the world. Jaguar and Rover announced to build VR centre in December, 2007 and Volkswagan Brazia formally announced its VR centre on June 15th, 2008.

VR is used in every area of automotive industry, such as styling, engineering, process, test, assembly and manufacturing. It not only provides freer working environment to engineers, but also refines and improves the development process. The promotion and application of VR will trigger a revolution to automotive technology.

3 Application Status of Virtual Reality in Chinese Automotive Industry

The automotive companies in China began to set up VR certres in about 2004, such as PATAC, FAW, HF, SVW, JAC; whereas some other companies including SAIC, Chang' an, Chery, etc., are planning to build VR centres or develope their VR technology . Figure 2 shows the VR centre in PATAC.

Most applications of VR in Chinese automotive companies are mainly in design engineering and manufacturing, including styling, assembly, CAE visualization, process visualization, etc.

The computer systems used in these VR centers include SGI super computer which is in the early VR system and PC cluster based on UNIX. But now they are

Fig. 2 The virtual reality centre in PATAC

moving toward PC cluster based on Windows while computer performance increases. Sometimes, only a workstation is competent. The reason is that VR system must be integrated into the whole vehicle development process. Windows and PC is the mainstream in the automotive industry.

Powerwall is the only projector mode adopted in Chinese VR centers at present, with the screen size as 7.2 m×2.4 m. It is enough for full size car. Few companies are using the CAVE, whose strength lies in giving a better immersive environment. That is more valuable to interior design. However, both GM and Opel design centers abandoned their CAVE, because they don't think immerse is critical to styling review and unsuitable for review by many people. However, as the data shows, GM engineering center keeps the CAVE. Actually, some other companies, for example, Renault uses the CAVE to assemble virtually.

The main software is the commercial software in these VR centers. Just for instance, for design review, Deltegen in RTT, Opus studio in Opticore (acquisitioned by Autodesk), Showcase in Autodesk are used; whereas Simens has applied VisMockup for virtual assembly and CEI uses EnSight for CAE visualization. What deserves to be mentioned is that Dassault Systems bought Virtool last year which is applied by Renault for virtual assembly. Recently DS is positively developing new 3D Via aiming at enhancing their capacity in virtual design and assembly review.

In addition, to improve the capability or upgrade the system, some VR centers are developing the VR with the cooperation of universities and academes. For example, PATAC did some study in VR technology cooperating with Shanghai Jiaotong University, Zhejiang University and Tongji University.

4 Virtual Reality Boosting Car Development

The competition pressure is forcing the car companies to develop more new products within a shorter time. However, it is not feasible any more to just relying on the traditional development process. Every car company is studying how to accelerate its development. GM started to implement its GVDP since 1990s mainly developing their math-based R & D capacity. Now it comes to its version 5.0, in which the computer-aided technology has been widely used. For example, in the past, there were several prototypes built during the process. Now, six kinds of virtual cars are used, replacing most of physical prototypes. GVDP is still constantly improving its development. One of its directions is to implement the VR, especially at the concept design phase.

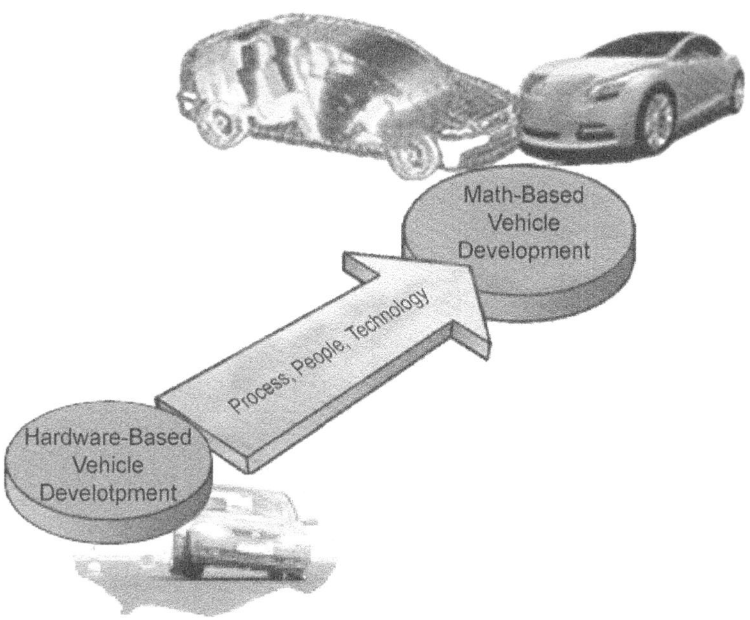

Fig. 3 Vehicle development: hardware-based to math-based

In China, VR is also showing its value in accelerating the car development, including styling, assembly, engineering and manufacturing [3].

In styling, with VR, the proposal can be reviewed by the realistic virtual model. All details can be evaluated to get the true feeling of the design. Thus the requirement of the physical model decreases, while the efficiency increases.

In engineering design, the capacity of VR providing the huge data loading and virtual assembly enables the engineers to sit together to review structure, assembly and

Fig. 4 Realistic vehicle modeling in virtual environment

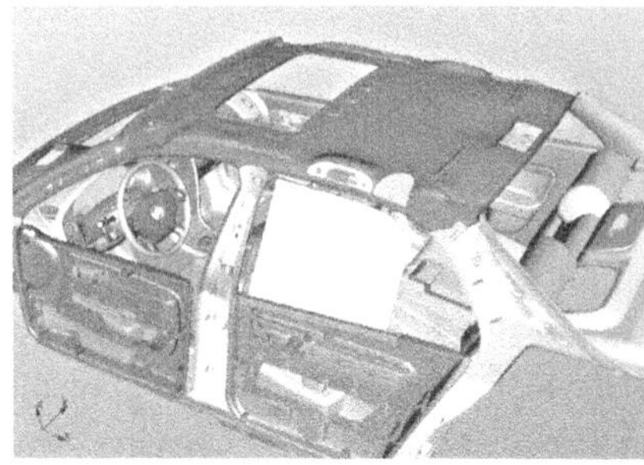

Fig. 5 Virtual assembly based on vehicle digital mockup

manufacturing feasibility. It is very useful to find the issues and make decisions. Even the workers can join in the discussion, which could avoid a lot of mistakes in low level. The prototype process gets faster.

In engineering analysis, the presentation of physical phenomena provided by VR makes the complex calculation results showed out in a more natural way. As a result, the un-expert engineers and designers etc. could fully understand and make use of the result. It is helpful to CAE to expend its utility.

In experimentation, with the introduction of VR, the test data can be dealt in better methods. By combining the physical and virtual properties, the test accelerates and some physical processes are simplified.

Fig. 6 Visualization of vehicle collision engineering analysis

Fig. 7 Vehicle test in virtual environment

In manufacturing, marketing and training, VR can find its application as well.

5 More VR Technologies Required by Chinese Automotive Industry

There are so many types of CAX software applied in the automotive industry, especially CAE. There are different types for different application. How to integrate CAX has become a very important issue for the companies to make better use of the work from each area and set up the perfect virtual development capacity. As a natural interactive tool of "people and virtual model", VR provides a good base to integrate knowledge which could be shared by the technicians in the whole chain. Consequently, VR is able to provide a big help to improve the development capability of companies.

Currently, VR is used in designing more than in manufacturing. The reason possibly is that manufacturing lacks the foundation data. Even in design, only display capability is used at most of time. Sometimes display is not enough, such as evaluating the acoustics and interacting with math model in real time.

Actually, VR can find its room in many areas, for example, ergonomics review based on virtual model, headlight illumination test, virtual crash test, noise evaluation, etc.

Additionally, collaboration is important for global development. How to make use of VR in collaboration is worth studying.

6 Approaches for VR Technology Better Satisfying the Needs of Automotive Industry

Even though VR is in development constantly, it still has some shortages. The perspective is an issue for Powerwall as there is only one position can get a correct perspective with the stereo glass while used for designing review by many people. As a result, everyone would get their own perspective which is different from each other. That is fatal to review. Therefore, the stereo has to be shutdown during review. That is why GM design center changed its powerwall to a big size screen some years ago. The stereo is regarded as worthless for review by a team.

The resolution ratio is another issue. Normally resolution ratio just reaches 1,200 pixel at maximum in the screen with the height of 2.4 meters, which is not enough to show the details clearly. This happened in one of our program, with the purpose of evaluating the driver's view based on the virtual model. But we could not clearly see the scene from the side mirror of the virtual car and got trouble in tracking the eyeball. We have applied the infrared camera to track the eyeball's position, but it is impossible in stereo model as eyeball is hided by shutter glass.

We also had a trial to con-review the design by VR model from different sites. However, each site has to keep a copy of the whole data, which could not be accepted by everybody.

Data rework is another issue. Most of VR need to rebuild the math model from CAD. If CAD data is changed, it needs engineer to update the virtual model manually, resulting in that the data can not be updated in time. This definitely limits the usage of VR. Now, the hardware rendering is attracting more attention. CAD data can be used for virtual model directly in a high level display card, which is the general trend.

The complex operation of software also limits the VR's usage. It is a tool for experts and the engineers or designers to follow the function set by the experts, as they can not operate the virtual model as the same as the physical model. Therefore, it is necessary to work out the software as simple as possible, to become a daily tool for engineer/design. Furthermore, it is quite important to popularize the VR.

In scientific calculation visualization, VR expands slowly. We made a program to replay CFD simulation in a VR environment. But CAE engineers didn't show their interest in it. They still regard VR as one of the post-process software.

Therefore, within the current situation, we can start our application of VR from the following aspects for making better use of VR.

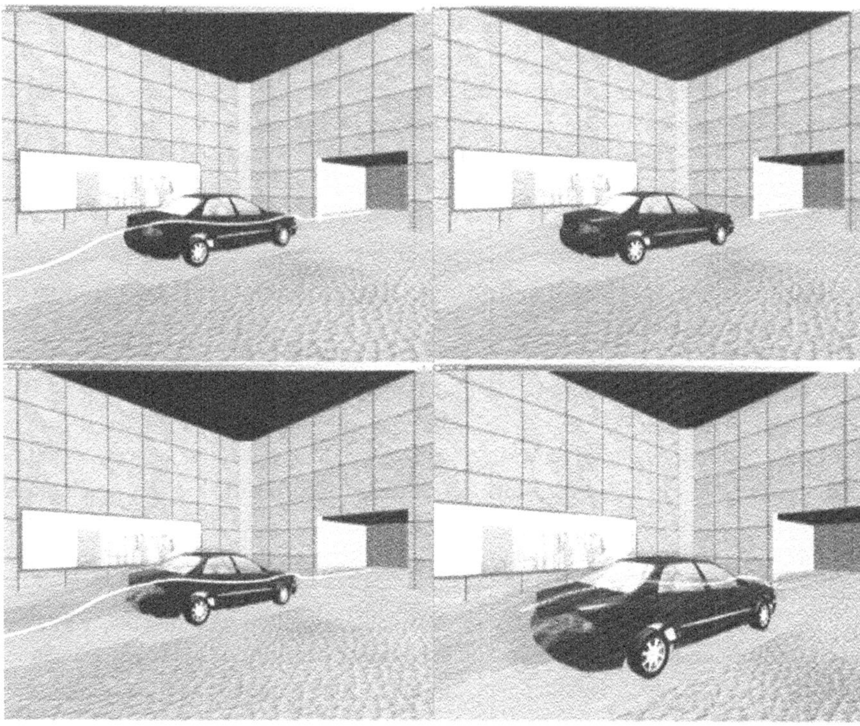

Fig. 8 Flow visualization of automobile CFD analysis

Firstly, enhance the integration of VR and CAX. VR must be seamlessly integrated into the whole development process with other CAX technology.

Secondly, start from the application which is not so timely expensive, such as ergonomics, driving simulation and virtual assembly. CAE application needs long time to calculate, which is not suitable for real time interaction.

In addition, AR, combining the virtual and physical model, is also very useful for automotive industry. However, because of the complexity, it is very hard to simulate everything by VR in automotive development. Making use of the physical model to simplify the simulation can give a more accurate result.

7 Conclusion

VR model is able to partially replace the traditional physical prototypes, as it is very helpful to accelerate the development, reduce the development cost, and especially to improve the capability of Chinese automotive industry. The math-based technology has been widely used in Chinese automotive industry. How to build up the whole virtual capability is crucial to the automotive companies now. It is great that automotive companies are laying store by VR. However, VR is still under development at the moment, both for enterprises and VR technology. The VR is just initialized in Chinese automotive industry and more effort is needed to better understand and make use of it. Besides, VR needs to be more improved to meet the requirements of automotive industry.

It is well known that Germany is the precursor in manufacturing. The VR is emphasized in the automotive companies in Germany. When Volkswagan launched its VR center in Brazia, the premier of Germany and the board chairman of Volkswagon attended the ceremony.This demonstrated how Germany emphasized the VR. At present, VR is used in the detailed works such as parts design, interior design, aerodynamics and crash simulation. According to the VR director of Volkswagon, VR is increasing the competitive capability. Germany uses VR to change the traditional industry[4]. It can be regarded as a model for the integration of digital economics and traditional economics. Germany is going ahead of the world in VR. It should be relative to German government's emphases on VR, which could be a reference for Chinese automotive industry.

Shanghai has put the advanced manufacturing on a very important strategy position and it is greatly supported by government to develop the advanced technology. As a high-end technology, VR should have a very good future within this environment.

References

[1] Wu Y, Zhang S J (2004) Application of virtual reality in the automobile (in Chinese). Automobile Science & Technology, 2004, issue 5

[2] Chen Y (2000) Germany uses VR to change traditional industry (in Chinese). Science & Technology Industry of China, 2000, issue 12, p 52

[3] VikiPedia. Virtual Engineering. http://en.wikipedia.org/wiki/Virtual_engineering, July, 2009

[4] Jiang M Q (2006) Digital technology and its application in car design. In: Proceedings of 15th Chinese Automotive Body Technology Forum, Society of Automotive Engineers of China, Automotive Body Committee, April 2006, pp 201–205

Author

Mr. Manqing Jiang joined SAIC in 1994. He started his career from Alias sculptor in SAIC technical center. He joined PATAC in 1997 since PATAC had been launched.

He has worked on several positions such as Packaging, program management, studio management, etc. In 2003, he became the digital senior manager of design department. His responsibilities include managing the digital design team to build math model, reviewing digital model, and releasing the math data. In 2004, he set up PATAC's VR center as the program manager. He is very interested in VR and pays a lot of attention to the math-based strategy in car design area and has written some articles and given some speeches about these topics. Besides, he has a very strong passion on new technology and has some collaboration with the universities, such as Shanghai JiaoTong University, Zhejiang University and Tongji University.

At present, he is the vice-president and secretary-general of transportation design committee of SIDA (Shanghai Industrial Design Association). His task is to support the communication and cooperation among industries as automobile, ship, aircraft and railway transportation and promote the development of industrial design in these industries.

Corresponding author

Mangqin Jiang(✉)
Design Department, Pan Asia Technical Automotive Center (PATAC)
No. 3999 Longdong Avenue,Pudong District,Shanghai,China, 201201
E-mail: Manqing_Jiang@patac.com.cn

Potentials of Innovative Technologies within the Scope of Complex Products and Services

Frank Thielemann
UNITY AG, Germany

Abstract

Today's requirements in product development, product quality and production process efficiency combined with the increasing product and production complexity represent a great challenge for staff members at all levels (from the assembly workers to the plant manager). The ultimate goal is the fine-tuning of the engineering and production process to perfectly fulfill customer orders and to keep the overall efficiency at high levels. In this context Virtual Reality and Mobile Computing seem to be the promising technologies to fill this gap. This paper describes the potentials of these innovative technologies and shows examples in different fields of application.

Keywords

Mobile computing, Product Development, Production Planning, Virtual Reality, Visualization

1 Introduction

Actually industrial companies act in a very difficult business environment, which can be characterized by an increasing dynamics of innovation and a reduction of the product lifetimes (as shown in Fig. 1). In the same amount the products and the production processes become more and more complex. This results in high demands on product engineering, production processes and the involved employees.

At the same time, enormous progresses have been made on computing performance, visualisation technologies, minimization of system components and hardware devices. Those developments benefited from the increasing game and entertainment industry. Here the rapid developments in the area of IT and computer hardware allow the production of very powerful computer hardware — in small structural shapes and for a low price. A

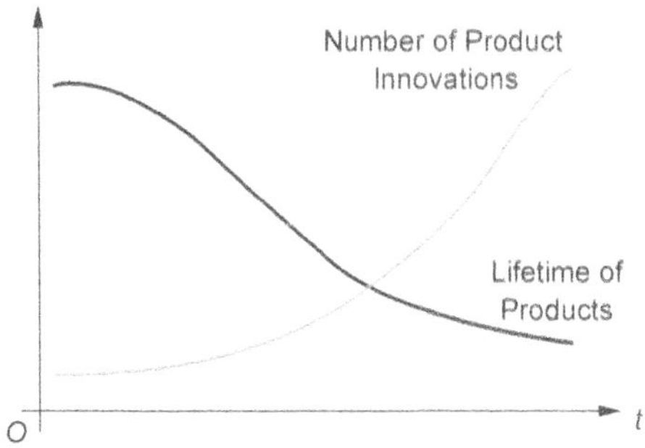

Fig. 1 Comparison of product lifetimes and innovation dynamics

mass market for 3D-hardware has been created. Each year multiple generations of powerful graphics cards have been developed and offered for the wide public for affordable prices.

The functionalities provided by up-to-date hardware devices can be used productively for fine-tuning of the engineering and production process to perfectly fulfill customer orders and keep the overall efficiency at high levels. Thus the technologies of Virtual Reality (VR) and Mobile Computing (MC) can be used productively in industry [1– 4].

2 Initial Business Case as Decision Basis

Initially –before the general project starts –a business case must be created. Basis for the business case is the monetary evaluation of the costs and the expected savings as well as the return on invest (ROI). This contains the analysis of the following aspects:

- **Development costs:** For the introduction of a new VR- or MC-application dedicated developments and investments have to be done regarding computer hardware, software, interfaces to existing IT-systems, etc. These costs have to be identified.
- **Process costs:** By introducing a new technology existing business processes can be modified, so that the expected workload can be increased. Furthermore it is possible that additional business processes be created (e.g. for the data generation, data supply, etc.). These costs must be included in the total costs, as well.
- **Process savings:** A reason for the introduction of a new technology is not the technology itself — it is the support and optimization of the existing business processes. This support and optimization can be estimated by the expected savings. Such savings can be the reduction of necessary business processes (e.g. checking of stocks by phone calls), the elimination of process cycles (e.g. design changes from engineering, which cannot be incorporated into the existing production line), the reduction of resources or the cancellation of dedicated process steps (e.g. the manual calculation of the current work load of a machine or an assembly line).
- **Not monetary assessable savings:** Some savings exist, which cannot be included as monetary savings into the business case. Such savings can be quality improvements or a higher level of planning reliability.

All these costs and savings must be considered within the creation of the initial business case (as shown in Fig. 2). Comparing the estimated costs with the expected benefits allows a reliable Go/No-Go decision.

The initial business case is used to decide whether to introduce the new technologies or to withdraw it.

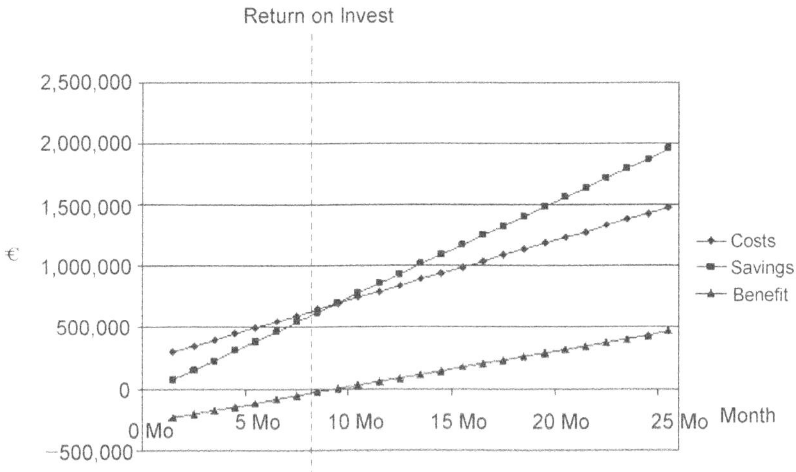

Fig. 2 Business case showing the costs, the savings and the calculated ROI

3 Key Success Factors for Introducing New Technologies

For a successful introduction of new technologies in a company dedicated success factors exist. If they are not considered adequately within the previous described project phases, the introduction will fail. These key success factors will be described as the following:

- **Data provision:** An efficient and effective data provision is a key success factor for the productive usage of VR and MC in industry. In a company a variety of data is generated, which potentially can be used by some applications (e.g. product designs, crash simulations, factory layouts, moulding and milling simulations, etc.). Due to the heterogeneity of the company's IT-systems and the non-central data storage and management, the process for data provision must be adjusted to the specific needs of the end users and the existing business processes.
- **Prevention of media breaks:** Today most companies design their products in 3D. But especially in machine and plant engineering the generated 3D-models will be converted into 2D within the subsequent business processes and productively used. Thus manufacturing and production, quality assurance, purchase, technical sales and service still use 2D-drawings for their daily work. The reasons are that certain information cannot be integrated easily into the 3D-models or no willingness exists in the departments to leave the chosen path. This leads to media breaks within information research or information processing. It is obvious that a media break in the information chain hinders, decelerates or reduces the quality of information research and processing.

- **Usage of standard data formats:** For the usage of geometry data in VR- or MC-applications native 3D-CAD-data must be tessellated in a first step. Therefore a variety of formats are available: JT (Jupiter Tessellation), XVL (extensible virtual world description language), X3D (Extensible 3D), etc. The availability of 3D-exchange-formats is manifold. But an established and wide accepted standard is not available. Thus companies are afraid that they bet on the wrong horse within the introduction of new technologies. Even the experiences with VRML have shown how hard it is to enforce a commonly accepted standard in this area.
- **Automatic data generation:** Today companies exist who use VR- or MC-applications productively within the product development or production planning processes. But mostly an automatic data generation is not available. The request for e.g. a "VR at the push of a button" is not always implemented completely. The reasons for this are the usual time intensive preparation of the data (assignment of materials and textures) as well as a not sufficient level of automation for the required steps. Furthermore the PDM-system managing the 3D-data must be set-up for an automatic data generation. Otherwise the PDM-system will be a bottleneck within the data generation process.
- **Bidirectional interface:** The data generated by intermediate formats are finally available with the VR- or MC-application. If changes or manipulations are made on the data (e.g. repositioning of a robot) these manipulations must be done on the original native 3D-data, as well. A bidirectional interface between the VR- and MC-applications and other IT-systems (e.g. ERP, 3D-CAD) is generally not available.
- **Central data management:** A central data management implemented within a PDM-system is not always available in the companies. Because local, self-developed databases are productively used by a variety of persons and departments. This data must be included into the data supply, as well. Furthermore it is mostly requested that not only geometric data, but also other data like component properties (e.g. used materials, production type) and the status of a spare part (e.g. released, not released, in work) should be provided within the VR- of MC-application. Such meta-data is not generally managed by a PDM-system, but stored in e.g. ERP- or CRM-systems.

4 Choosing the Right Technology

To make best use of the indubitable high potentials of VR and MC in industry, the introduction of the technology should not be the predominant issue. The technology is only one module, to support the existing business processes in an optimal way (as shown in Fig. 3).

Moreover the technology must follow the company internal business processes. So

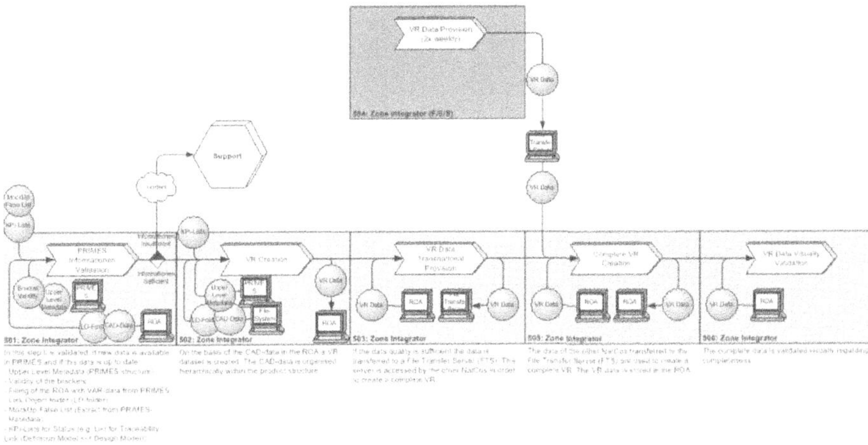

Fig. 3 Analysis of the business processes, the existing IT-systems and the required data

the following aspects must be considered:

(1) **Anchoring of technology in business processes:** During the introduction of a new technology it must be anchored in the existing business processes. Dedicated process steps must be implemented, which postulate explicitly the usage of this technology. This prevents the rudimentary usage of new technologies by the employees and therefore a weak exploitation of the offered potentials.

(2) **Functionality follows process:** The implemented application must follow the existing business processes. It must be prohibited that an introduction of a new technology leads to a significant modification of work steps so that existing processes are adopted to the technology. Otherwise this will lead to a low acceptance of the application and island solutions.

(3) **Definition of milestones for data provision:** Dedicated milestones must be defined where the required data must be available in a specified quality and maturity.

(4) **Early involvement of end users:** Only an early involvement of the real end users allows the identification of all requirements. This will lead to a higher acceptance of the implemented application.

(5) **Identification of existing bottlenecks:** For the integration of a new technology potential bottlenecks must be identified. Such bottlenecks base mostly on IT-limits (system limits due to data amount, missing bandwidth of the network, etc.) or process-related constraints (process dependencies, time limits for data processing, etc.).

In conclusion the focus must be a process oriented derivation of requirements (as shown in Fig. 4). Based on the analysis of the business processes, the existing IT-systems and the data requirements must be defined on the following levels: process requirements, functional requirements, organizational requirements and integration requirements.

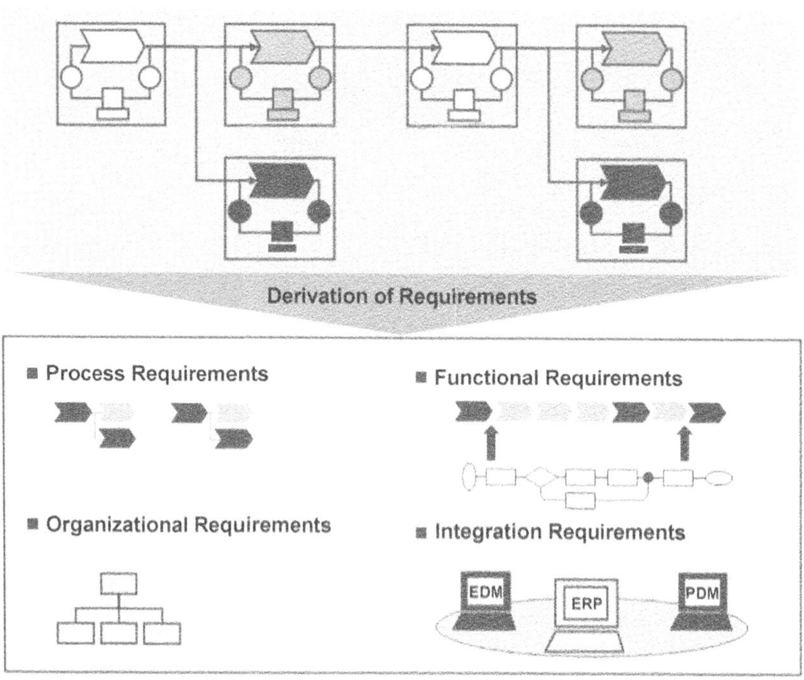

Fig. 4 Introduction of new IT-systems — from processes to qualified requirements

5 Exemplary Usage of Innovative Technologies within the Scope of Complex Products and Services

5.1 Virtual Reality for Training of Complex Products

The complexity of current products rises more and more. This complexity has to be managed by the company in general and by the employees in particular. Due to the high dynamics of innovation the products have a significant alteration rate. These changes require a high degree of flexibility and qualification from the employees. Here usually classical training methods meet their limits. The usually high complex facts cannot be communicated to the employees in an understandable way. Here innovative training concepts are necessary, which present in an understandable manner the product properties, its functionalities and further facts. The VR-technology can meet these requirements.

An example for this is the visualisation of hydraulic pumps. In this context, beneath the complex mechanical behaviour the hydraulic processes play an important role. The pumps can only be explained and described correctly in connection with the interaction of mechanical and hydraulic components. Beneath this the developed VR-application visualises the functional dependencies of the pump components, provides exploded views and offers the possibility to enrich the VR-scenes with circuit diagrams, technical drawings and corresponding textual explanations (as shown in Fig. 5). Additionally the user can interact with the application in order to adjust the settings of the pump (e.g. modification of the current control and change of the pivoting angle).

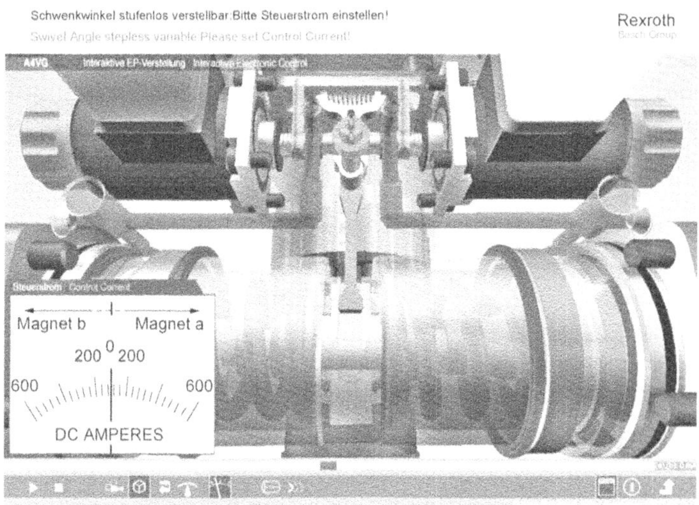

Fig. 5 VR-visualisation of a virtual adjustable pump and its current control. By modifying the current control the adjustment of the pump is visualised (source: Bosch Rexroth)

5.2 Virtual Reality for Process Visualisation

In case of failures or malfunctions of facilities and machines a correct troubleshooting is a key success factor. It allows only a short downtime and a quick recommissioning. This aspect becomes even more important if the facility or machine is very complex. Initially a first troubleshooting can be made by the operators on-site. If the problem cannot be solved by them, the machine and plant manufacturer must be consulted.

For the first troubleshooting for a light exposure machine a VR-application has been developed. It visualises the current status of the machine in real-time (as shown in Fig. 6). Therefore the operating data of the real machine is read

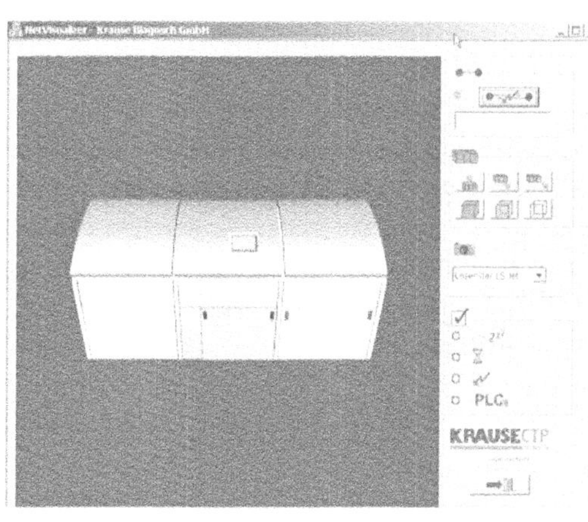

Fig. 6 VR-visualisation of a virtual (source: Krause CTP)

out continuously and analysed. And the virtual model of the light exposure machine is adjusted accordingly. This VR-visualisation allows an easy casual research and a quick bug fixing on-site.

5.3 Mobile Computing for Preventive Maintenance

The avoiding or minimization of machine downtimes is crucial for all producing companies. Therefore a preventive maintenance is usually performed. Such a regular servicing and the exchange of wearing parts (if necessary) before a malfunction of the machine occurs will minimize the downtime of machines. For such maintenance tasks a mobile computing application has been developed (as shown in Fig. 7).

Fig. 7 Documentation of performed steps for preventive maintenance on a mobile device

This MC-application supports the current preventive maintenance activities. It shows the single maintenance steps, which have to be performed by the staff, and documents these steps, the results, the maintained parts and the exchanged components. For a smooth spare parts supply the system is wireless connected to the existing SAP-system. The customers benefit is the quality intensification of preventive maintenance, the reduction of the training effort required for new employees and the IT-supported analysis of the weak points of the machines based on the on-site documented results.

6 Summary

The high potentials of current VR- and MC-technologies have been shown for a variety of industrial fields of application since several years. Nevertheless the predicted wide usage of these technologies is still not completely visible.

Only a reduction of the personal effort required for the rollout and the operation of a VR- or MC-system coupled with a standardisation, IT- and process-integration leads to a successful and well established usage of these innovative technologies in industry. But the introduction of these technologies should not be the main goal. The technology is only a module, which must support the existing business processes. Thus the technology must follow the current business processes. In order to achieve a wide usage, the explicit anchoring of the technology within the business processes is an important issue.

A company is a network of IT-systems, processes and persons. It can also be described as a triangle (as shown in Fig. 8). The dependencies of persons, processes and systems within a company must always be balanced. All the three areas should be enhanced adequately — an imbalance has to be avoided. Here the final goal is to enlarge the whole triangle — not only the part of IT-systems or the technologies.

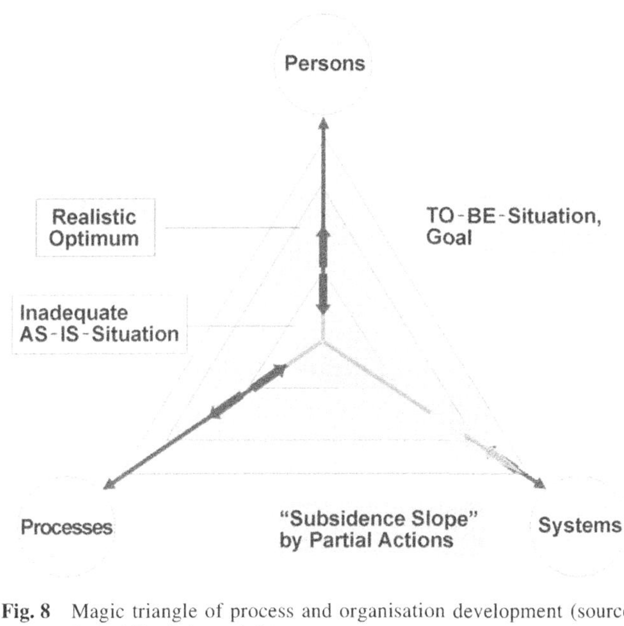

Fig. 8 Magic triangle of process and organisation development (source: Stein-hilberSchwer AG.)

References

[1] Göbel M (2000) Innovation oder Integration? Ein Jahrzehnt Virtual Reality. In: Der GMD-Spiegel

[2] Altmann A (1998) Cyberfabriken: Virtuelle Realiät im industriellen Einsatz. In: c`t, Heft 15

[3] Bauer C (1996) Nutzenorientierter Einsatz von Virtual Reality im Unternehmen. In: Computerwoche Verlag GmbH

[4] Kimberley W (2000) Process Innovation with Virtual Reality. In: Automotive Engineer 25

Author

Dr.-Ing. Frank Thielemann, born in 1969. Having specialized in computer-integrated production (now Product Development), from 1996 to 1998 he worked as both a researcher and a consultant for the Heinz Nixdorf Institute at Paderborn University, where in 1998 he completed his doctorate under Professor Gausemeier. The focus of the research and consultancy work that he carried out for a large number of enterprises, was on the reorganization of product development processes. Today Frank Thielemann is a member of the management board of UNITY, a technology-oriented consultancy company specializing in strategies, processes, technologies and systems. UNITY supports its clients throughout their product development processes, continuously ensuring high productivity and successful implementation of changes processes. As a member of the management board, Frank Thielemann is responsible for the areas of Lean Development, Mechatronics Management, Digital Factory and Innovation Management.

Corresponding author

Frank Thielemann(✉)
UNITY AG, Lindberghring 1, 33142 Bueren, Germany
E-mail: frank.thielemann@unity.de

From Immersive Engineering to Selling and Teaching

Martin Zimmermann, Andreas Wierse
VISENSO GmbH, Stuttgart, Germany

Abstract

The development of immersive applications is demonstrated based on the COVISE software: started as a tool for the analysis of numerical simulations it has been extended to the whole field of engineering. Based on the realisation that the communication of complex three-dimensional data is not limited to the development teams of enterprises, COVISE is today also used in the field of sales and marketing. And it now finds its way into universities and schools in order to make it easier to understand three-dimensional phenomena.

Keywords

Virtual Reality, Sales and Marketing, Engineering, Education

1 Introduction

COVISE stands for Collaborative Visualisation and Simulation Environment and has been developed at the High Performance Computing Center at Stuttgart University (HLRS) [1,2]. VISENSO stands for Visual Engineering Solutions and the VISENSO GmbH is a Spin-Off from the HLRS that makes COVISE commercially available and continues its development (in cooperation with the HLRS). Based on a modular approach COVISE today supports a large range of data interfaces in order to import the customer data; visualisation modules then prepare this data for the display in the virtual environment.

The COvise Virtual Environment Renderer COVER handles the visualisation and the interaction in the 3D world in a wide range of virtual environments from CAVEs to Powerwalls, curved screens, simple back projections to head mounted devices and autostereoscopic displays. The applications cover a wide field, including aerospace, architecture, automotive, machinery, medical engineering as well as research and development areas related to numerical simulations.

2 The Technology

We define Virtual Reality in this context as follows: the two core aspects are immersion and interaction (see also [3]).

2.1 Immersion

Immersion, meaning "being embedded" in the 3D representation, is achieved by using the stereoscopic display technology. In the real world humans use their two eyes to safely determine distances in the near field area. In the digital world the two images, that show the perspective from the respective eyes, are created by usually one or two computers; two projectors equipped with polarising filters then display these images on a suitable surface and in combination with polarising glasses each image reaches the corresponding eye. The brain then combines the synthetically created images in the same way that it handles those from the real world and creates the three-dimensional impression.

Today in most cases projectors are used that allow several people to work together. A typical size is a screen width of 2.5 to 3 meters. There are smaller, more compact installations, down to mobile setups that can even be transported in a normal car and are installed in a few minutes almost everywhere. On the high-end side the so-called CAVE (Cave Automatic Virtual Environment) allows the almost perfect immersion; a cube that consists of projection surfaces so that the viewer inside sees computer generated images wherever he looks. The cost of such an installation however is significant, given the number of computers and projectors (6 to 12) and the fact that a three-storied building with a reasonably large foot print is needed in order to project on all sides of the cube.

Therefore most installations today use the one channel back projection systems mentioned above (one display channel consisting of an image for the left and one for the right eye). They deliver most of the time the optimal price/performance ratio.

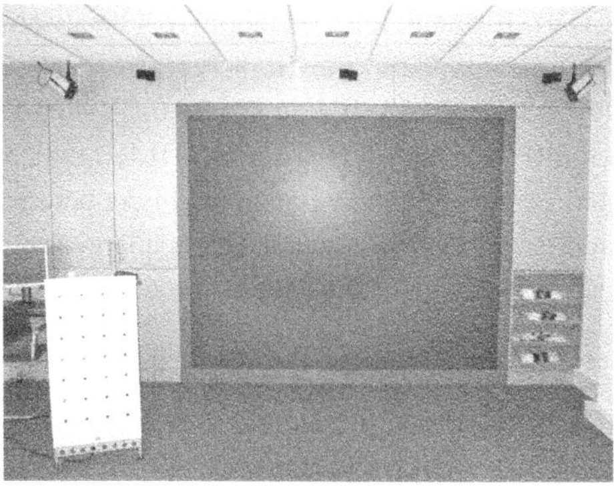

Fig. 1 typical single-channel VR installation (at VDC Fellbach)

2.2 Interaction

A special kind of 3D mouse is used for the 3D interaction. The best systems today use an optical tracking system for the position definition. By using markers (as shown in Fig. 2) a computer can compute, based on the knowledge of the camera position the coordinates of the interaction device and the glasses of the master user.

This allows the software to compute the exact perspective for this viewer position. Due to the fact that this happens several times per seconds (at least 20) the user can freely move around; the permanent adaptation of the image creates the impression that he can easily move in the virtual world and for example walk around the virtual product.

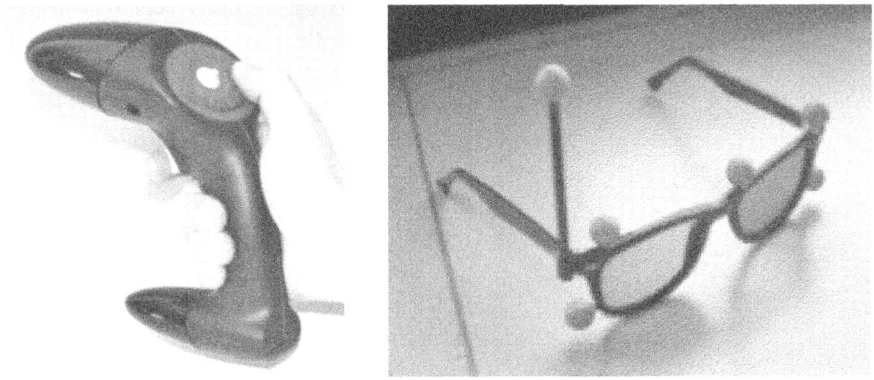

Fig. 2 Interaction device and eyeglasses with markers

Both aspects together, immersion and interaction resp. stereo display and 3D tracking, allow a very easy handling of even complex visualisations.

2.3 Software

In order to bring the virtual environment to life of course a software is needed that supports immersion as well as interaction. The COVISE software introduced above has been developed to do exactly this, especially strong in the field of the simulation of physical properties.

The interaction when working with simulation results (e.g. the positioning of cutting surfaces or switching the visibility of parts) is performed directly in the three dimensional world, utilizing the 3D mouse. Emphasis has been put on one side on the availability of all relevant data interfaces for simulation and geometry data (e.g. LS-Dyna, Abaqus, CAD, etc.); on the other side the interaction in the VR world has been kept simple, so that everyone can use this tool without long lasting training sessions.

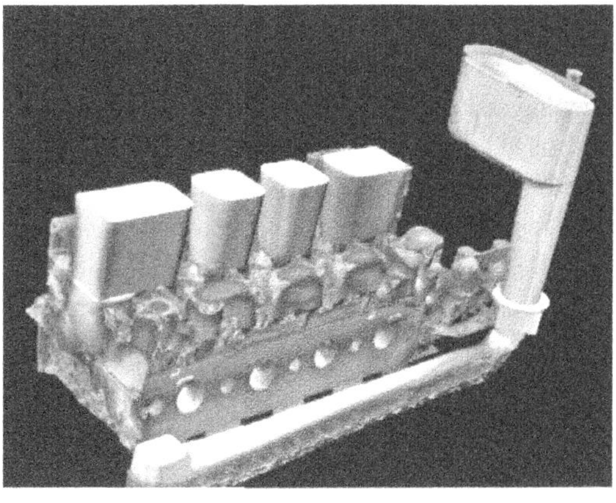

Fig. 3 COVISE visualisation of a casting simulation

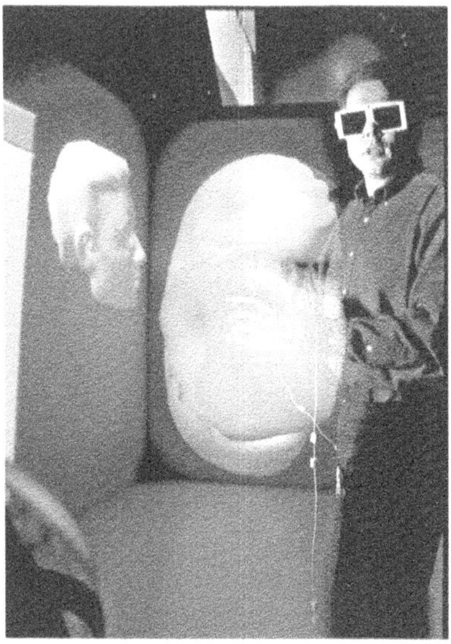

Fig. 4 Immersive engineering visualisation in a CAVE

3 Immersive Engineering

Here we find the classical development engineers, designers or numerical simulation engineers, who discuss the content of their work either with colleagues or their management. The Virtual Environment is in this context a communication means to ease the exchange between different specialists: each engineer can present the results of his or her work in a way, that makes it easy for the other specialists to set their work in the correct relation to that.

This is a well-established technology, which is part of the official development process. Increasingly enterprises install virtual quality gates, i.e. no longer only real world prototypes are used to assure the quality of the progress, but many aspects can already be verified in the virtual world. Decisions are based on completely virtual models, saving these enterprises a significant amount of money and time [4].

4 Immersive Selling

The visualisation in a Virtual Reality Environment however not only helps the specialists in their mutual understanding, but of course also eases the communication from the specialist towards the customer. The display of complex properties in a simple manner in combination with the possibility to show things in the virtual world, that cannot be seen in the real world (e.g. to look into the squeezing machine during the pressing action), is a very powerful tool for sales and marketing.

In addition a 3D visualisation itself is already a strong attraction point at a trade show or fair, for example for passers-by. This is perfected by the possibility to include animations and even complete company presentations into such a Virtual Reality presentation.

We already have performed a significant number of such presentations in the last years, including:

- the presentation of a new eyeglass lens for Zeiss at a tradeshow for opticians (OPTI);
- the presentation of a new ski for Salomon at a sports tradeshow (ISPO);
- the presentation of a new generation of a squeezing machine for Dieffenbacher at a sheet metal working technology exhibition (Euroblech)(see Fig. 5);
- the presentation of the manufacturing branch of Miele at the Hannover trade

show (see Fig. 6);
- the presentation of the advanced technical functionality of the new E-Class for Mercedes-Benz [5].

Fig. 5 Dieffenbacher VR-installation at Euroblech

Fig. 6 Miele booth at Hannover Fair

5 Immersive Teaching

The next level has been reached, by moving the Virtual Reality technology out of its strong engineering environment into the field of knowledge transfer in the context of school and university. If the technology is suitable to communicate complex three-dimensional content from specialists to customers, it is not a big step to also use it for the communication of three-dimensional phenomena from teachers to their pupils.

VISENSO has made a first step into this fields by providing a Virtual Reality environment to the Science House at the Europa-Park in Rust, Germany, several years ago (in cooperation with the Virtual Dimension Center, TZ St. Georgen). Here young visitors can interactively explore data sets from research and industry with the means of Immersion and 3D-interaction: flow fields in aero-dynamics, the flow of water in a river, the flow of the air through the nose, …

The next step has focussed on topics that are directly part of the school curriculum. Together with teachers we have identified topics that are explicitly three-dimensional and therefore difficult to understand for a significant part of the students. Starting with the theory behind magnetic fields in physics, this has recently been extended into the human ear: the way that the sound travels and the occurrences inside the ear until the sound is actually perceived are visualised interactively. This allows the student not only to see (like in a movie), but to also play with the scene, thus allowing a much more thorough understanding than before [6].

Fig. 7 Immersive teaching application: magnetism

It is remarkable that the application of this technology is especially helpful for those students that until now could not imagine what is going on. By bringing these processes to their mind in a more realistic way, it is made much easier for them to understand them; in addition the good students also benefit, because the virtual exhibit allows them to play around and thus makes it easier to derive new knowledge.

6 Technology

An important role for all three areas plays the technical progress in the field of display and interaction devices: while in the past Virtual Reality installations required investment in at least 6 – digit Euro or Dollar figures, today devices with a reasonable quality are available for as little as 2.000 Euro or Dollar. And even below that amount first devices are available that will help to spread the usage of Virtual Reality tremendously.

Here we would like to briefly present two typical examples:

First, the availability of 3D TV sets. Inspired by the movie industry, that is increasingly producing 3D content (already one third of the US movie theatres is equipped with stereo projection!), the TV manufacturers are preparing themselves for a 3D home TV market. Companies like Samsung make TV sets available that can display stereo content at a reasonable resolution. This opens up the usage of Virtual Reality for small and medium sized companies as well as for educational institutions. Even in the smallest office a place can be found for a screen that can of course also be used for example for PowerPoint presentations. And the investment for such a system can even be born by schools.

The second interesting technology can be seen in the field of head mounted devices: the Cinemizer from Zeiss can bring 3D visualisation immediately to the viewer for less than 400 Euro (see Fig. 8). Initially only planned for iPod-owners that want to watch their movies on the road, the Zeiss engineers were smart enough to add stereo capabilities to the Cinemizer. Therefore it is now possible to also view 3D content with it. First trials with classes were very successful: the students enjoyed the new technology and were able to see familiar topics in a completely new way, while the teachers were very fond of the increased concentration of the pupils on the topic.

These developments on the display side will be rounded off by progress on the interaction side as well: While 3D interaction devices like for example optical tracking systems were not available below 10.000 Euro, the appearance of mass market game like the WII from Nintendo will lead to a whole new bunch of interaction devices at a cost of less than 100 Euro (see Fig. 9). Of course these devices today cannot fully replace a

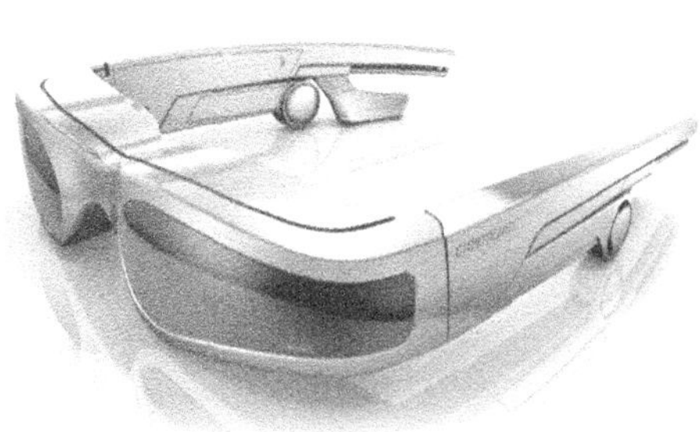

Fig. 8 Zeiss Cinemizer

full blown optical tracking system. But if we look at PC graphics cards today and how the graphics world was less than ten years ago: If the development of interaction devices is only half as fast as that of graphics cards we will have interaction devices better than those available today in less than five years.

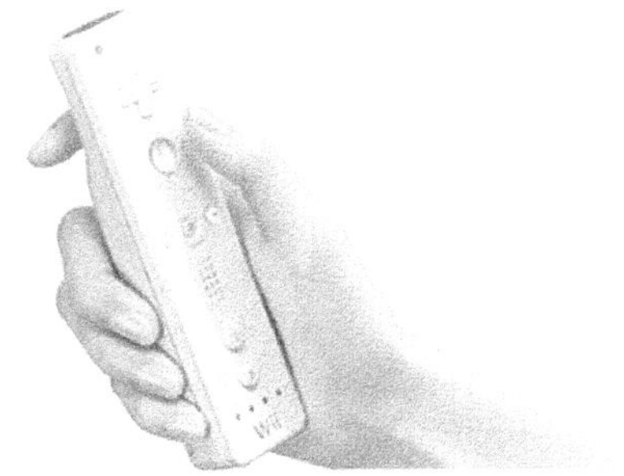

Fig. 9 Nintendo WII interaction device

7 Outlook

Our COVISE software has been on the market for more than ten years now. It started on graphics supercomputers that cost millions of Euros and was available only to a small number of specialists in very large companies or research institutions. Today we have reduced the necessary investment to less than hundred thousand Euros and even SMEs start to use our technology.

But the development that is going on today will make Virtual Reality available to each and every household. COVISE, due to its modularity and its openness is an ideal tool to handle the variety of application fields from special industrial problems via sales and marketing down to the interactive needs of edutainment for schools as well as for the general public.

References

[1] Wierse A, Lang U, Rühle R (1994) A System Architecture for Data-oriented Visualization, Database Issues for Data Visualization, Lecture Notes in Computer Science, Vol 871, pp 148 – 159, Lee, Grinstein (eds), Springer

[2] Wössner U, Rantzau D, Rainer D (1998) Interactive Simulation Steering in VR and Handling of Large Datasets. In: Proceedings of the IEEE Youth Forum in Computer Science and Engineering (YUFORIC) Germany '98, Stuttgart, 16 – 18 June 1998

[3] Durlach N, Mavor A (1995) Virtual Reality: Scientific and Technological Challenges, National Academy Press, Washington, D.C., 1995

[4] Wierse A (2008) Virtuelle Realität für die Produktion: Motivation und Beispiele. In: Neuere Entwicklungen in der Blechumformung, pp 357 – 362, Liewald (ed), Stuttgart, 2008

[5] Paul A (2009) Mercedes-Benz zeigt E-Klasse in Stereo 3D, http://www.visenso.de/fileadmin/PDFs/2009/09 – 09 – 01_Mercedes-Benz_zeigt_E-Klasse_in_Stereo_3D.pdf. Accessed 13 Jan 2010

[6] http://www.cyber-classroom.de. Accessed 13 Jan 2010

Author

Martin Zimmermann is founder and managing director of VISENSO, responsible for Sales and Marketing. He has an economical background, starting in the early nineties as CFO at Meta Finanz, a company owned by Allianz. Starting from 1998 he supported Plenum, an IT-consulting company, during their going public and then founded working ideas in 1999, an internet portal that he sold in the year 2000, before he entered the field of the Virtual Reality.

Corresponding author

Martin Zimmermann(✉)
VISENSO GmbH, Nobelstraße 15, 70569 Stuttgart, Germany
E-mail: mz@visenso.de

Index

The manufacturer's authorised representative in the EU is Springer Nature Customer Service Centre GmbH, Europaplatz 3, 69115 Heidelberg, Germany. If you have any concerns regarding our products, please contact ProductSafety@springernature.com

Printed and bound by CPI Group (UK) Ltd, Croydon, CR0 4YY
31/07/2026
02182903-0001